Environmental Regulation and the Allocation of Coal

Alan M. Schlottmann

The Praeger Special Studies program—
utilizing the most modern and efficient book
production techniques and a selective
worldwide distribution network—makes
available to the academic, government, and
business communities significant, timely
research in U.S. and international economic, social, and political development.

Environmental Regulation and the Allocation of Coal
A Regional Analysis

PRAEGER SPECIAL STUDIES IN U.S. ECONOMIC, SOCIAL, AND POLITICAL ISSUES

Praeger Publishers   New York   London

Library of Congress Cataloging in Publication Data

Schlottman, Alan M   1949-
    Environmental regulation and the allocation of coal.

    (Praeger special studies in U.S. economic, social,
and political issues)
    Bibliography: p. 137
    1. Coal mines and mining—Environmental aspects—
United States. 2. Electric power-plants—Environmental
aspects—United States. 3. Environmental policy—
United States. 4. Linear programming. I. Title.
TD195.C58S34   333.8'2   76-56841
ISBN 0-275-24090-8

PRAEGER PUBLISHERS
200 Park Avenue, New York, N.Y. 10017, U.S.A.

Published in the United States of America in 1977
by Praeger Publishers, Inc.

Printed in the United States of America

ACKNOWLEDGMENTS

I am indebted to Charles L. Leven and Edward Greenberg, both of Washington University, for their helpful comments and advice concerning an original draft, and to Richard Gordon of Pennsylvania State University for comments on a final draft. Harold J. Barnett's assistance in suggesting several topics to be considered in the analysis is gratefully acknowledged.

The assistance of Haskell Wald, U.S. Federal Power Commission (FPC) during the initial data collection effort cannot be overemphasized. Dr. Wald's aid was quite important in obtaining preliminary copies of the FPC Form 423 data which were essential to this study. Furthermore, he has either personally filled my requests for additional information or has put me in contact with the proper FPC personnel, particularly Alex Gakner and Norton Savage. They, and members of Dr. Wald's staff not known to me, have shown great willingness to aid in my research, and I am thankful for their assistance.

T. Reed Scollon of the U.S. Bureau of Mines was helpful in suggesting methodology and providing data for the early formulation of the regional mining-cost estimates. Leonard Westerstrom, U.S. Bureau of Mines, has been a constant dependable source of data and his help has minimized mistakes and errors in this study occurring from insufficient or old data.

Will Stockton, Vice President, Peabody Coal Company, during several conversations provided insights into the industry's view of coal's potential and the effects of public policy on the industry, particularly on expansion. He provided partial copies of reports which were otherwise difficult to secure. He also directed me to Robert Young of Peabody's Traffic Department who was most helpful in providing data for the construction of the transport matrix used in this study.

Recent conversations with the chairmen of computer science departments at two large universities support my finding that the operating manuals for mathematical programming packages can be rather incomprehensible, particularly for the class of users they are designed to address. I am most grateful to Lawrence Abrams, University of California, Santa Cruz, who gave me computational help when I was experiencing great difficulty in setting up the basic model. Like Dr. Abrams, I found that I actually needed a working model before I could learn the programming manual and his help saved a great deal of time in completing this study.

Most of this manuscript was written while I was employed at the Appalachian Resources Project at The University of Tennessee and its support in the form of typing, editing, computer budget, etc., is gratefully acknowledged. The continued support and encouragement of Robert Bohm and John Moore during this time was invaluable. This work was supported in part by Grant S1A72-03525 from the National Science Foundation/RANN to the Appalachian Resources Project.

LIST OF TABLES

LIST OF FIGURES

The major use of coal today is in the production of electricity and the fortunes of the coal industry have become tied to steam electric generation plants. There are two distinct classes of coal, high grade coal with metallurgical properties and so-called "steam coal," whose major value lies in the production of heat or steam. Because electric utilities provide such a great demand for this latter grade of coal, it is often referred to as "steam electric coal."

In 1973, utilities consumed 85 percent of the steam coal produced, while all other manufacturing, mining, and retail users comprised only 15 percent of the market.[1] The entire metallurgical coal market was only 25 percent as large as the electric utilities' demand for steam coal.[2] The electric utility demand for coal has increased 6.4 percent annually and is the only sector of the coal market which has experienced any growth during this period.[3] Coal, of course, is the predominant fossil fuel input in conventional steam electric generation, producing 56 percent of such electricity in 1973, while oil and natural gas generated 21 and 23 percent, respectively.[4]

As consumption of coal by electric utilities has increased, so has concern over the environmental effects of both coal mining and its use as a fuel. The effects on land use of uncovering coal resources buried under overlying strata are at issue as well as the effect on air quality of the emission of various pollutants during coal combustion.

Coal reserves in the United States are found in three general areas. The Eastern area consists of the long Appalachian fields stretching southward from Pennsylvania into northern Alabama. The Midwestern, or Eastern interior, fields are concentrated mainly in Illinois, Indiana, and Western Kentucky. The Western fields, where the coal industry is a comparatively new development, comprise those deposits lying west of the Mississippi.

The characteristics of coal vary from one area to another and within the same area. The environmental issues raised in land use and air quality for the coal industry tend to accentuate the importance of these regional differences. Our main purpose in this study is to consider the regional implications of environmental control policies as they relate to the electric utility market for coal.

The following two chapters discuss the environmental concerns raised in the mining and utilization of coal. Chapter 1 deals with land use issues arising from mining. Chapter 2 is concerned with the air-quality issues involved in the utility industry itself and in other coal use.

In Chapter 3 the spatial linear programming model used in evaluating the effects of proposed environmental policies on regional mining activities is introduced. The basic solution to the model is presented in Chapter 4. An evaluation of a moderate sulfur emissions standard is given in Chapter 5.

The interaction of land use and air quality proposals is seen in Chapters 6 and 7 where the empirical results of the programming analysis are evaluated. Chapters 8 and 9 evaluate the prospects for Western and Eastern coal fields as they relate to the model's results, including a discussion of the significance of these results for the competitive position of coal in relation to that of other fossil fuels. The coal fields of the Midwest are discussed in Chapter 9 (Eastern Coal).

NOTES

1. Derived from data in National Coal Association, *Bituminous Coal Data 1973* (Washington, D.C.: National Coal Association, 1973), p. 77.

2. Ibid.

3. As computed by Richard L. Gordon in *U.S. Coal and the Electric Power Industry* (Baltimore: Johns Hopkins University Press, 1975), p. 3.

4. Federal Power Commission, *Monthly Report of Cost and Quality of Fuels for Steam Electric Plants* (Washington, D.C.: Government Printing Office, December 1973).

1

THE LAND IMPACTS
OF COAL MINING

A SHORT PRIMER ON COAL-MINING METHODS

The extraction of coal can be classified into two general methods, underground mining and surface mining.[1] In underground mining, tunnels are dug into the underground seam and, within this space, special equipment is used to withdraw the coal without disturbing the overlying strata. This general mining method is termed "room and pillar." The tunnels, referred to as "rooms," are driven into the seam parallel to each other. Rooms range generally from 14 to 20 feet in width, and the roof is stabilized by "pillars" of coal as well as artificial supports. As the mine is worked out, the coal pillars are often removed.

Longwall mining is a European technique in which the coal face of a seam is sheared off by mechanical planes or plows. Only jack supports are used for roof stabilization, and, since the jacks follow the machines, the roof collapses behind the machines as the seam is penetrated. In the United States, longwall mining accounts for only a small percentage of production, 2.6 percent in 1972.[2] Though recovery rates from underground seams vary considerably, ranging from 29 to 91 percent, the average of 57 percent is relatively low.

In surface mining the overlying stratum is removed, exposing the underlying coal seam. Explosives are often used to break up the stratum and make it more easily transportable. The capital equipment, such as power shovels, bulldozers, and draglines needed to remove the overlying rock and dirt, is usually the largest of its kind. The shovels may be up to 220 feet tall with the capacity to remove 270 tons or 180 cubic yards of material every 50 seconds. The largest dragline has a boom of 310 feet and a bucket that holds 220 cubic yards.

Surface mining can be divided into two main types, area and contour mining. Area mining is employed where the terrain is relatively flat or generally rolling in nature. The coal is removed and overburden deposited in long trenches.

In the steep terrain of mountainous areas contour mining is used. Here the overburden is removed from above the outcrop of the coal bed and deposited at or over the edge of the slope. Auger mining, where machines drill horizontally into the seam, is used in conjunction with contour mining on steep slopes to increase the output. Since there is no need to leave coal pillars in the seam to act as roof supports or to anchor artificial supports as in underground mining, the average recovery rates for surface mining are high, approaching the better recovery rates of underground mining.

The relative importance of surface and underground mining on a regional basis will be discussed in Chapter 3 as well as the relative extraction costs of the two methods.

LAND-IMPACT PROBLEMS

The land impact of coal mining can create significant difficulties.[3] Because of their diverse methods of operation, underground and surface mining pose different problems. Surface mining physically disturbs only the land that is mined, but it may have harmful effects on neighboring lands as well. Unreclaimed land is eliminated from alternative uses. Water quality can be affected significantly through sedimentation and silting of streams or acid mine drainage.

Underground mining may also have an adverse effect on land. Waste piles left exposed at the surface and improper mine sealing can contribute to acid mine drainage. Removing an underground coal seam increases the chance of land subsidence. However, the most controversial effects of underground mines considered in the following discussion have been on miner health and safety.

SURFACE MINING

The environmental problems associated with surface mining and nonreclamation of land can be severe. As of 1972, 2 million acres of disturbed surface land had not been reclaimed,[4] and 20,000 miles of highwalls remain exposed, with sheer cuts resembling those cliffs that occur in contour mining. The water quality of thousands of miles of streams and lakes has been severely degraded.[5] In addition to these economic effects, the habitat of wildlife has been disturbed and the beauty of the land has been impaired.

Studies have indicated that the frequency and intensity of flooding increase with the level of mining activity, but no attempt has been made to quantify these damages.[6] Current research at the Appalachian Resources Project (ARP) is directed toward filling this gap.[7] As shown in Table 1.1, their results indicate that damages from flooding are generally greater in areas where there is surface mining. The mean level of damages in surface mining counties is almost double

TABLE 1.1

Mean Flood-Damage Levels in Ohio, Tennessee, and Virginia, 1962-72

	Mean Values in	
Variable	Stripped Counties	Nonstripped Counties
Death	0.5098	0.3099
Injuries	16.6667	9.4930
Houses destroyed	0.2353	1.3239
Houses—minor damage	110.3725	54.6056
Houses—major damage	5.7647	9.0563
Farm houses destroyed	0.2941	0.2535
Farm houses—minor damage	0.0000	0.3521
Farm houses—major damage	0.7255	1.0000
Families suffering loss	148.4902	94.4084
Families assisted by Red Cross	42.6470	35.9718
Individuals given Red Cross care	254.7059	218.5915
Mobile homes destroyed	3.4314	1.3662

Source: F. K. Schmidt-Bleek et al., "Benefit-Cost Evaluation of Strip Mining in Appalachia," Appalachian Resources Project (Knoxville, Tennessee: The University of Tennessee, 1974), p. 5. Values from a sample consisting of 99 percent of all floods occurring in the eastern United States between 1962 and 1972 (of an intensity great enough to require action by the Red Cross).

the level in nonsurface mining counties for the five important categories of death, injuries, number of houses with minor damage, families suffering loss, and mobile home damage. The mean levels of nonstripped counties exceed those of stripped counties only in the categories of houses destroyed and major house damage. When all types of residences are considered, however, the mean of residences destroyed is larger in surface mining counties. The rate of water runoff and the problems it generates for waterways appear to be greater in areas where land has been disturbed. Though the data developed are for northern Appalachia, it is generally apparent that surface mines can increase pressure on natural drainage capacity.

When vegetation is removed from the land, erosion of the surface may result. The large spoil piles left by surface mining consist of highly erodible material, and, if the spoil material is steeply graded or is piled on steep slopes, the risk of erosion is increased. Landslides create a problem on steep terrain.

Surface water runoff can contaminate receiving streams through sedimentation. The relationship between these factors and mining-slope angle is positively correlated.

Another general problem associated with surface runoff and mining wastes is acid mine drainage. Not all water draining from mines is affected by acid and related pollutants, but a major share of the economic damage from mine drainage is acidity related. A 1969 report estimates that 90 percent of acid water pollution is related to coal mining.[8] The source of this pollution is pyrite, a compound of iron and sulfur frequently encountered in and around coal deposits. When pyrite is exposed to air and water, it is oxidized. Water runoff carries the pyrite into surrounding streams where the iron oxide component precipitates to the bottom, visibly polluting the stream.[9] Since sulfuric acid is toxic to water plant life and animals, the affected streams can become biologically dead.

When strip-mine land is not reclaimed it is removed from present as well as future uses. Although agriculture and livestock grazing are not feasible on unreclaimed land, reclaimed land is often devoted to pasture and renewed farming use. Well-planned reclamations, such as those in which lakes are formed, can often enhance the aesthetic and recreational appeal of an area, while unreclaimed land is only an eyesore.[10]

EXTERNAL DAMAGES FROM PRIVATE MARKETS

The major effort to control the environmental harm of surface mining has been through government regulation. The U.S. House of Representatives considered a total of 17 bills dealing with surface mining in 1974. Government intervention is necessary because surface mine prices do not seem to reflect the true social costs of production. The economist, accustomed to dealing with market prices, must be concerned with this problem as well. Most environmental problems are concerned with this divergence between private and public costs of production.

The general equilibrium model of economic theory is the theoretical basis for allowing markets to determine price and quantity relationships. To begin with, there is a set of consumers with given preferences and an initial allocation of resources. Producers are assumed to have production relationships which are well defined and behaved. Generally, if the producers have convex production sets and consumers exhibit convex consumption sets, an equilibrium system of prices and quantities is determined. The introduction of long planning horizons and uncertainty does not prevent a general equilibrium solution from being attained, but markets must exist for all goods, services, and contingencies over time. The solution is viewed as efficient, in a Pareto sense, in that the condition of any one individual or group in society cannot be improved unless that of some other individual is made worse. Given our initial preferences and assumed allocations, the prices in the system are "correct signals of social value."

The criticism of surface mining is based on the assumption that something has gone wrong with the pricing system. One possible problem is that all the production costs are not explicitly incurred in the production of the product. Our general equilibrium model assumes that all production costs are internalized, that is, that external costs or spillovers do not occur. Damage to land or water quality by surface mining, should be reflected as production costs, but most external effects of this type are not accounted for in market transactions. Thus, prices do not reflect the true social costs of production.

Private and social costs of production have diverged, and markets responding to private costs only will misallocate resources. In such a situation, government control through mining reclamation regulations may be necessary. Government action could attempt to directly internalize the external damages as costs of surface mining by taking damage payments from the offending firms and distributing them to those suffering the damages. Reclamation laws and other such regulations attempt to simulate the desired outcome by placing restrictions on those production activities which have been identified as the source of external damages.

By comparing the costs of resources employed in reclamation to the benefits of abatement, the economist would determine the degree to which surface mining should be subject to reclamation requirements. The optimal level of reclamation would occur when the marginal value of resources used in eliminating these external effects equals the marginal benefits of abatement. The practical problem is that the costs of abating these external damages must be correctly determined for the range of reclamation control under consideration. Correctly assessing reclamation costs and evaluating dollar benefits of surface mine control are not easy tasks.

One study attempted to quantify the benefits of reclamation in dollar terms for the Kentucky state statutes, which contain rigorous requirements for regrading and revegetation.[11] The data manipulation and estimates are fairly elaborate, if unsophisticated. However, the general idea is to discount damages from five major sources to their present worth, while assuming that natural forces tend to reduce damages from each source over time. The five sources considered are acid mine drainage, silting damage, landslide costs, costs to adjacent land surface and aesthetic damage. Richard Gordon, who attempted to revise and update the methodology used, concludes that the benefits of surface mine reclamation are measurable at approximately $50-150 per acre.[12] The data set used in such estimates is often inadequate, and, in order to make more precise estimates, the ARP gathered extensive information on the direct external effects of surface mining. The flood-damage differentials discussed earlier are an example of this. Preliminary work by Robert Bohm has indicated that direct external damages on a per-ton basis for surface-mined coal in Appalachia are 80 cents to $1.00 per ton. For an "average" disturbed acre, these damages can represent $1,600-2,000 per acre.[13] This per-acre level of damage is quite high when we note that land in surface-mined counties can sell for as low as $125 per acre. Assuming these figures are correct, they represent the maximum level of reclamation costs that would be imposed on surface mining. Proposals which would require higher expenditures per acre must assume that there are large benefits which are not being measured.

In any case, benefit measurement could be understated since it is often public goods that are affected by surface-mining externalities. Public goods are those that are available to individuals in a society, whether or not they ever use them, and the provision of public goods to one person does not exclude other members of society from using them. The classic examples of public goods are police and fire protection, but clean air and water are also in this category.

Private markets find it almost impossible to efficiently provide these services,[14] because of the difficulty in determining and coordinating the individual's and society's preference for such goods. Individuals have rational reasons for not revealing or for distorting their preferences for public goods. An individual's desire for a good may decrease if he must pay for it, rather than if society provides it. This is the well-known free rider problem.

Within reasonable limits, there is no additional cost of providing a unit of a public good to an additional user. Private firms could not afford to provide services where the efficient allocation of resources does not exclude users and implies a zero price. Many members of society might be willing to pay for the restoration of natural habitats and scenic views, even if they never were going to visit them.[15] But as a public good, the cost of providing such options to an additional person is zero. Thus, while the restoration of land and natural habitats through adequate reclamation may provide a wide range of goods and services, there is neither a market to provide them nor a market price to evaluate them.[16] The role of government reclamation control in this case is not to emphasize the direct external costs of surface mine production, but to prevent a permanent conversion of natural land into unreclaimed mining sites.

Even if we assume that all benefits of reclamation are correctly measured, considering public goods and production externalities, there are still other factors which could influence optimal reclamation levels. Some advocates of reclamation have argued that the removal of surface-mined land from future alternative uses is a "cost" of surface mining. Initially we might disagree with this position. If a landowner could earn higher returns in alternative uses, he would not grant mining rights and allow surface mining in the first place. Even if these rights were sold several years before mining began, they could be repurchased if more profitable alternative opportunities arose. If unreclaimed land lowers future land returns, this calculation has been figured in the landowner's discounting process. This argument is not correct from society's viewpoint in the "real" world.[17]

In the general equilibrium model considered earlier, we noted that the introduction of uncertainty did not affect the result that the economy was ex ante efficient. Yet we needed to assume that a complete system of insurance markets and contingent commodity markets existed. These insurance markets would cover all possible events, such as whether or not stack-gas scrubbing is completely successful by 1980, etc. Not only would a general equilibrium solution determine present levels of commodities, but also contingent ones at each point in time.

In the real world, a full set of such future and insurance markets is not available. The main problem is that a divergence between private and social rates of discount in investment can occur. One group of economists, including Kenneth Arrow and William Nordhaus, believes that the discount rate in the United States is too high, with the positive rate differential a result of risk and taxes. Suppose owners of resources bear risks due to uncertain price changes, changes in technology, expectations, etc. This group of economists has argued that these are not social risks. Since they can be widely spread over the population, their effects on output are extremely small compared to average income.[18] If so, then the private discount rate will be above the appropriate social discount rate. If the interest rate before taxes is the social discount rate, the presence of capital taxes will also make the equilibrium rate too high for private decisions. As Nordhaus shows, distortion of the interest rate changes price paths to such a degree that resources are consumed too fast for their most efficient allocation. According to Nordhaus, a further result of distorted interest rates is that the private planning horizon is too short, due to the absence of future markets. As an alternative society may provide resources to reclaim surface mines and develop future land uses over a longer planning horizon. The shorter planning horizon for an individual may not take into account sufficient future earnings on reclaimed land.

To help eliminate surface-mining problems, it has been proposed that mines be required to reclaim disturbed land. There are state regulations at present in most major coal-producing areas.[19] In the Appalachian areas, all states have statutory requirements regulating strip mining. Certain revegetation is required, and the overburden deposited at the edge of the mining bench or downslope must be smoothed over. These minimum requirements in Appalachia have been copied by other coal-producing areas. The Midwestern states of Illinois and Indiana have similar laws.

In order to provide uniform requirements, however, federal legislation has been proposed and is pending. The House of Representatives considered 17 bills in 1974, and H.R. 11500 received preliminary passage in September 1974. The House version, when reconciled with a slightly different Senate bill, was passed but vetoed by President Ford. The final legislation will almost certainly be stringent. A discussion of the main legislative points being considered on the Eastern and Western fields appears in Chapters 8 and 9. An evaluation of imposing reclamation costs is attempted in Chapter 6.

RECLAMATION AND THE ORPHAN MINE

It has been shown that the environmental damages from unreclaimed land can be severe and are the major source of external effects from coal mining. These problems, usually from abandoned mines, are difficult to correct through common law litigation, since mine operators often claim financial insolvency after the mining operation has ceased.

Many of the early estimates for reclaiming abandoned or orphan mines were based on minimal efforts to offset the water pollution from sedimentation and acid mine drainage.[20] More recent work has been concerned with the costs of reducing the hazard of landslides, returning the land to productive uses, and reestablishing a natural appearance. Since landslides, erosion, and slumping spread the spoil over a wider area, earth-moving costs on these orphan mines are greater than on those in operation. In Appalachia, vegetation which has grown spontaneously on the spoil must be removed before regrading. Using a mining operation's capital equipment for reclamation is cheaper than transporting it back again to the orphan mine.

The Council on Environmental Quality (CEQ) study summarizes the costs of reclamation for previously mined lands. A study in Pennsylvania concerned with restoring a previously mined area to original contour found an earth-moving cost of $2,770 per acre (1965 dollars). For comparison, they calculated a cost of $923 per acre to restore current mines to original contour.[21] A more recent study for the ARC indicates that regrading and backfilling costs in 21 orphan-mine operations over 1967-71 were correlated only with the degree of spoil dispersion on the downslope.

The longer that reclamation is delayed, the larger the general dispersion of the spoil and the greater the reclamation costs. The future use of restored lands, that is, the degree to which the contour must be restabilized and the vegetative cover must be reestablished, determines the extent of reclamation required. For example, the ARC study indicates that the costs of terrace backfilling average only 57 percent of those in contour backfilling. In West Virginia, their analysis revealed a per-acre cost of $526 and $923 for terrace and contour backfilling, respectively.[22]

In order to fund the reclamation of previously mined areas, it has been proposed that federal legislation levy a fee on all coal, both surface and underground. Some of the revenue would be used to treat acid mine drainage at abandoned underground mines, but the majority would be used to restore land at orphan surface mines. This tax is to be set at a uniform 35 cents per ton on all coal mined in the United States.[23] This would add from 1.3 to 1.8 cents per million Btu to the cost of Eastern and Midwestern coal and approximately 1.8 to 2.2 cents per million Btu to the use of Western coal.

There are two interesting issues in the imposition of such a tax. First, is it appropriate to levy a tax on the current coal industry to pay for the environmental degradation of past mine operators? Unless current coal companies are the same as those which operated in the past, in which case the tax would be a type of time-lagged reclamation expenditure, there is no economic argument that the entire coal industry today is responsible for yesterday's degradation. Unless the same parent company is involved, the relatively new Western producers have not contributed to Appalachia's problems. However, 40 percent of the funds raised would be used in the state where they are collected. In states with few abandoned mines, as in the West, the funds collected would be used partly for public services such as schools or roads. Since there are a large number of smaller independent operators, in addition to the major coal companies, the burden of past responsibility does not necessarily carry over to the present.*

Second, the coal fee is locationally neutral, and its imposition, across all regions and types of mining, does not change the patterns of regional coal use and distribution in the model, except for the effects of cost increases on interfuel substitution.[24] For a tax of 35 cents, this would represent a negligible increase of less than two cents per million Btu. Such a fixed tax, whether at 35 cents per ton or any other level, does not cause any change in relative locational advantage or mining method differentials. This would raise annually about $210 million. For comparison, the disturbed acreage of orphaned strip mines has been estimated at between 2 and 4 million acres. The required reclamation costs for these lands could run as high as $8 to $12 billion.[25] Since all reclamation efforts may not be completely successful, some reseeding and replanting might be necessary in the future. This would be financed through the coal-fee bill.

We have noted that 78 percent of acid drainage stems from inactive mines, with abandoned underground mines causing two-thirds of that amount. The ARC has estimated that an expenditure of $6.6 billion would be required to eliminate acid mine drainage problems nationally.[26] Gordon estimates that Appalachia's outlays would be about $5 billion of the total.[27]

A study prepared for the commission has estimated measurable benefits of abatement at less than $30 million a year.[28] As Gordon points out, at a 10-percent rate, $500 million in benefits a year would be required to repay Appalachia's control expenditures.[29] The benefits reported by the ARC and by

---

*The large companies in the Midwest have generally carried on adequate reclamation projects as indicated in a conversation with Mr. Will Stockton, Vice President, Peabody Coal Company.

Gordon are not the only answers, but they suggest that the main problem with such acid mine drainage is the pollution of waterways. Again, however, the nature of public goods and the time horizon for decisions to restore natural areas are relevant in any benefit study. These studies suggest a cost of approximately $4.4 billion to eliminate acid mine drainage in inactive underground mines. Thus, if full restoration of all mining areas and complete elimination of all environmental damage from inactive mines were attempted through a coal fund, the per-ton tax would be with us for years.

SLOPE-ANGLE CONSIDERATIONS
IN APPALACHIAN PRODUCTION

As mentioned earlier, the slope at which surface mining takes place can be an important variable in environmental damages, with erosion, landslides, and the sedimentation of streams increasing as the slope angle increases. When slopes exceed 20 degrees, stabilization of spoil material becomes very difficult, and this instability has two major impacts. First, because rainfall not only adds to the weight of the spoil but lubricates the spoil material and the surface on which it is located, the potential for landslide is severe. Some landslides have occurred even after reclamation. Second, sedimentation of streams and lakes, even at considerable distances from the original area, results from surface erosion. There is inadequate control of surface runoff on 98 percent of surface mined land in Appalachia.[30] At present, 70 percent of all surface mining in Appalachia occurs on slopes greater than 15 degrees and 51 percent on slopes greater than 20 degrees. Outside Appalachia, less than 10 percent of all surface mining takes place on slopes greater than 15 degrees.[31]

A serious proposal is to ban all surface mine production above a certain slope angle. Because it appears to be the critical angle for spoil stabilization, 20 degrees is the limitation most discussed. Senator Henry M. Jackson has stated that "to a major extent the debate over federal surface mining legislation has narrowed to the question of what the impact of imposing various forms of slope degree limitations would be."[32] The CEQ estimates that a limit of 20 degrees would reduce Appalachian strip mining by 51 percent, total U.S. strip mining by 20 percent, and total coal output by 10 percent. Therefore, a ban would have a major impact, and we will return to a discussion of it in Chapter 5.

UNDERGROUND MINING

Though underground mining does not tear away the surface material over the coal seam, it can cause land subsidence. Removal of the coal seam removes support for overlying strata. With decreased support the roof can bend and crumble, causing settling, and depressions and potholes at the surface. The effects are site specific. Some lands, such as those used for agriculture or forests, may be only slightly affected by subsidence. Because of the increased chance of damage to buildings and other structures, the greater the proximity to urban areas, the more the subsidence becomes a matter of concern. In Maryland and Ohio, subsidence cases have been relatively few, small in effect and in remote areas. In West Virginia, subsidence cases are also in remote areas but are larger and more numerous. Pennsylvania, where anthracite mining is fairly close to urban areas, has suffered most.[33]

Most subsidence damage is from older, abandoned mines. The threat of subsidence decreases as the depth of the mine increases, but early underground mines worked shallow seams since surface-mining equipment was not advanced enough to remove the overburden economically. Subsidence might also be controlled by supplementing coal pillars with artificial supports, but these were seldom used in early mines.

The two other methods of controlling subsidence discussed are partial extraction and backfilling. Partial extraction simply leaves much more of the seam in place for support than merely pillars. Backfilling refers to methods of supporting the cavities by filling them with mine wastes, crushed rock and other material. Most backfilling techniques are impracticably expensive and are believed to be only 50 percent effective in reducing surface subsidence.[34] As Gordon notes, "Apparently, there is no fully satisfactory way permanently to avoid subsidence."[35]

Refuse piles outside of underground mines are often heavily laden with pyrite material which oxidizes to create acid mine drainage. Attempts to cut off oxygen by sealing the mines are extremely difficult. Water treatment has been stressed as a means of control. The technique of lime neutralization, which reduces the acidity and removes most of the dissolved minerals associated with it, is most widely used.

Most acid mine drainage comes from abandoned mines where there was little attempt at treatment. Inactive mines are the source of 78 percent of Appalachia's acid drainage, with inactive underground mines accounting for two-thirds of this amount.[36] Current mining attempts to treat acid mine drainage and is not responsible for most spillover effects. Cost data on treatment of current underground mines are summarized in a study of *Resources for the Future*.[37] For the most difficult cases in the worst site locations, the highest published estimate is 75 cents a ton. Other studies indicated a cost of around 10 to 20 cents per ton. The best source is a more recent study on water treatment costs at 22 coal mines. Excluding the three lowest results due to their favorable locations, the average control cost was still only 13 cents per ton. The median cost was even lower, 8 cents per ton. This report also indicates that treatment to abate water pollution is customary at many underground mines in the Appalachian region. Of the Appalachian streams and water affected with

TABLE 1.2

Bituminous Underground Mine Closing Data, 1969-72
(thousand short tons)

| | Reason for Closing | | | |
| | 1969 Act[a] | | Other[d] | |
Region	Mines	Tons[c]	Mines	Tons
Appalachia	138	2,885	1,401	53,998
Midwest	7	423	10	4,244
All other states[b]	9	47	20	684
United States rated annual production				
Plus 250	0	0	23	12,221
100-250	7	1,004	93	14,063
50-100	10	770	111	7,883
50	137	1,581	1,204	24,759
Total	154	3,335	1,431	58,926
Present distribution of U.S. total				
Plus 250	0.0	0.0	1.5	19.5
100-250	0.4	1.6	5.9	22.6
50-100	0.7	1.4	7.0	12.6
50	8.6	2.4	75.9	39.9
Total	9.7	5.4	90.3	94.6

[a]1969 Federal Coal Mine Health and Safety Act.
[b]Arizona, Arkansas, Missouri, New Mexico, Oklahoma, Utah, and Wyoming.
[c]Rated annual production.
[d]Other categories reported were: (1) market conditions or inferior product, (2) personnel problems, (3) economic conditions, (4) adverse mining conditions, (5) mine worked out, (6) unknown or not available.
Source: Aggregated from Tables 1 and 2, United States Bureau of Mines *Bituminous Coal and Lignite Mine Openings and Closings in the Continental United States, 1970, 1971, 1972* (Washington, D.C.: Government Printing Office, 1973), pp. 2-6.

acid drainage three-fourths are in Pennsylvania and northern West Virginia. Pennsylvania has legislative requirements for treating acid mine drainage. Outside of Pennsylvania the laws dealing with water pollution from coal mining and acid drainage are less rigorous.[38] Coal deposits in southern Appalachia are surrounded by strata containing alkaline materials which neutralize acid. Therefore, acid drainage is more of a problem of surface runoff and the wastes of unreclaimed surface mining.

The main issue associated with underground mining has been the general problem of miner health and safety. In 1970, the disabling frequency rate was 3.8 times the average for all industry. The severity rate, total number of man days lost per million man hours worked, was 8.4 times the national average and the largest in the country.[39] Underground miners are subject to pneumoconiosis, a disease that results from long periods of breathing respirable coal dust. In the late 1960s approximately 100,000 miners were afflicted with this disease, better known as the "black lung" disease. The mine explosion and fire in 1968, which claimed 78 lives in Farmington, West Virginia, resulted in the 1969 Federal Coal Mine Health and Safety Act, subjecting underground mines to respirable dust standards and safety requirements.

The effects of this act on the underground mines in the Eastern coal fields is not entirely clear. Costs of operation were affected since the addition of some

FIGURE 1.1

Fatality Rates—All Coal Mines—1950-73

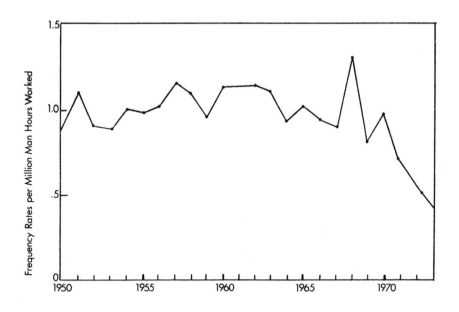

Source: United States Department of Interior, *Administration of the Federal Coal Mine Health and Safety Act* (Washington, D.C.: Goverment Printing Office, 1974), p. 7.

FIGURE 1.2

Progress in Meeting the 2.0 mg/m³ Respirable Dust Level, January 1971 to December 1973

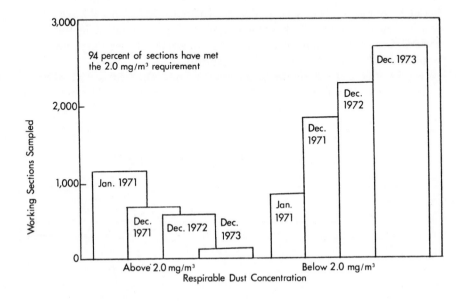

Source: United States Department of Interior, *Administration of the Federal Coal Mine Health and Safety Act* (Washington, D.C.: Government Printing Office, 1974), p. 14.

new equipment and the modification of some existing equipment were needed along with associated manpower requirements. Some industry estimates of the cost of the legislation run as high as $1.50 a ton.[40] This figure was based on the cost of retrofitting older mines, and estimates might be lower for newer mines or for all-new operations with engineering measures to meet the requirements. It has been suggested that the crucial impact was not the new mine requirements themselves but rather that they were passed at a time when other adverse forces were operating.[41] The addition of new workers and labor unrest during this time raised costs as productivity decreased, and estimates of the act's costs could well have been biased by including the effects of these other factors. A cost figure of 60 cents a ton, which was derived in a study by the Charles River Associates, is generally considered to be a more accurate estimate of the effects of the act.[42]

Table 1.2 shows the closing data of underground mines from 1969 to 1972. As shown, the majority of mines which reported closing from the effects of the act were small operations. Smaller operators, faced with the cost of replacing disallowed although usable and undepreciated equipment and constrained by labor and lending institutions, may well have decided to shut down. With uncertainty over the impact of sulfur emissions on future production, the prospect of additional investment appears to have had a "last straw" effect on small mines in Pennsylvania, Ohio, etc.

By 1973 it appeared that the majority of coal mines, including the smaller operations, had installed the necessary new equipment and modified their mining practices and older machinery to comply with the health and safety standards of the act.[43] For the third consecutive year, fatalities were the lowest in the statistical history of the industry. The frequency rate reached its record low of 0.45 per million man hours. Figure 1.1 shows the change in the rate from 1950. By comparison, the Peabody Coal Company, with 80 percent of its production from surface mines, had a frequency rate of 0.94 over the period from 1968 to 1971.[44]

Since exposure to concentrated dust and possible pneumoconiosis poses the greatest threat to the health of underground miners, the law set a standard level for respirable dust. Figure 1.2 shows the progress in attaining this standard from 1971 to 1973. The fact that 94 percent of the sampled sections have met the standard clearly shows that progress in this area has been achieved. Federal inspections, both planned and unannounced, have risen steadily and may be one reason for the improved record of the coal industry in this area. In 1973, for example, inspections increased by more than 50 percent over 1972, which had already increased by 50 percent over 1971 inspection levels. A tentative conclusion may be that the main effects of the act have already been felt in the coal industry and that underground operators particularly have adapted their operations to this new institutional structure.

NOTES

1. The discussion here is mainly based on National Coal Association, *Bituminous Coal Facts 1970* (Washington, D.C.: National Coal Association, 1971), pp. 12-17, and United States Department of Interior, *Surface Mining and Our Environment* (Washington, D.C.: Government Printing Office, 1967), pp. 33-49.

2. National Coal Association, *Bituminous Coal Data 1973* (Washington, D.C.: National Coal Association, 1974), p. 49.

3. These are discussed in Council on Environmental Quality, *Coal Surface Mining and Reclamation* (Washington, D.C.: Government Printing Office, 1973).

4. Ibid., p. 1.

5. Ibid.

6. Herbert A. Howard, "External Diseconomies of Bituminous Coal Surface Mining— A Case Study of Eastern Kentucky." (Unpublished doctoral dissertation, Indiana University,

Bloomington, Indiana, 1969). Also see C. R. Collier et al., *Influences of Strip Mining on the Hydrologic Environment of Parts of Beaver Creek Basin, Kentucky, 1955-66*, Department of Interior (Washington, D.C.: Government Printing Office, 1970).

7. The following is based on F. K. Schmidt-Bleek et al., "Benefit-Cost Evaluation of Strip Mining in Appalachia," Appalachian Resources Project (Knoxville, Tennessee: The University of Tennessee, 1974).

8. Appalachian Regional Commission, *Acid Mine Drainage in Appalachia*, H. R. Doc. No. 90-180, 91st Congress, 1st Session, XXIV, 1969. Contains appendices and summaries of environmental studies.

9. Based on the discussion in E. F. Goldberg et al., *Legal Problems of Coal Mine Reclamation*, Environmental Protection Agency (Washington, D.C.: Government Printing Office, 1972).

10. See Mined Land Conservation Conference, *What About Strip Mining* (Washington, D.C.: Mined Land Conservation Conference, 1964).

11. Howard, op. cit.

12. Richard Gordon, "Environmental Impacts of Energy Production and Use," "Energy Supply Project" (unpublished manuscript on file, Resources for the Future, Washington, D.C., 1973), pp. 39-40.

13. Robert Bohm et al., *Benefits and Costs of Surface Coal Mining in Appalachia*, Appalachian Resources Project (Knoxville, Tennessee: The University of Tennessee, 1974).

14. Paul Samuelson, *The Collected Scientific Papers of Paul Samuelson*, Vol. 2 (Cambridge: Massachusetts Institute of Technology Press, 1966), pp. 1223-39.

15. This concept of option demand is considered by Robert Spore, "The Economic Problem of Coal Surface Mining," *Environmental Affairs* 2, no. 4 (June 1973): 585-693.

16. For a case study of such a conflict between natural habitat and surface mining, see Robert Spore et al., "Opportunity Costs of Land Use: The Case of Coal Surface Mining," in *Energy-Demand, Conservation, and Institutional Problems* (Cambridge: Massachusetts Institute of Technology Press, 1973).

17. The discussion here is based on William Nordhaus, "The Allocation of Energy Reserves" (paper presented at Brookings Panel on Economic Activity, Washington, D.C., November 15, 1973).

18. For a discussion of these concepts, see Kenneth Arrow, *Essays in the Theory of Risk Bearing* (Chicago: Markham Publishing Company, 1971), particularly Chapter 11.

19. A review of these is contained in Council on Environmental Quality, op. cit., Appendix F.

20. Ibid., Appendix D, contains a discussion of reclamation costs for previously surface-mined areas.

21. Ibid., p. 97.

22. Ibid., p. 98.

23. St. Louis *Post Dispatch*, September 25, 1974, p. 2.

24. Costs which do not vary with deliveries have no effect on the minimum cost solution or shipment patterns in these linear programming models. In particular, costs which are uniform across regions have this effect. See James Henderson, "A Short Run Model of the Coal Industry," *Review of Economics and Statistics* 37 (1955): 337.

25. Schmidt-Bleek et al., op. cit., p. 24.

26. Appalachian Regional Commission, *Acid Mine Drainage in Appalachia*, H.R. Doc. No. 91-180, 91st Congress, 1st Session, 1969, p. 13.

27. Gordon, op. cit., p. 45.

28. Appalachian Regional Commission, op. cit., p. 12.

29. Gordon, op. cit., p. 45.

30. Goldberg, op. cit., p. 39.

31. Council on Environmental Quality, op. cit., p. 6.

32. Council on Environmental Quality, op. cit., p. vii.

33. These relative state effects are from Goldberg, op. cit., pp. 46-47.

34. Ibid., p. 49.

35. Gordon, op. cit., p. 36.

36. Appalachian Regional Commission, op. cit., p. 41.

37. See Gordon, op. cit., pp. 47-48, for a review of the following studies.

38. The statutes are reviewed in Goldberg et al., op. cit., pp. 83-90.

39. *Accident Facts 1971* (Washington, D.C.: National Safety Council, 1971), p. 35.

40. Congressional Research Service, *Factors Affecting the Use of Coal in Present and Future Energy Markets* (Washington, D.C.: Government Printing Office, 1973), p. 23.

41. Gordon, op. cit., p. 36.

42. Charles River Associates, *The Economic Impact of Public Policy on the Appalachian Coal Industry and the Regional Economy* (Cambridge: Charles River Associates, 1973), p. 183.

43. The following discussion and figures, except as noted, are based on United States Department of Interior, *The Administration of the Federal Coal Mine Health and Safety Act* (Washington, D.C.: Government Printing Office, 1974), Chapter I, pp. 7-15.

44. United States Congress, House Committee on Interior and Insular Affairs, Hearings, *Regulation of Surface Mining,* Part I (Washington, D.C.: Government Printing Office, 1973), p. 788.

2

ELECTRIC UTILITIES
AND THE ENVIRONMENT

The role of electricity as a major energy source is well established. Total output of electricity in the United States more than doubled in the 10-year period from 1961 to 1971.[1] Though energy consumption is strongly correlated to GNP, more efficient utilization brought a secular fall in the ratio between World War I and 1967.[2] The ratio of energy to GNP has increased since 1967 by about 10 percent.[3] Whether this is a reversal of a secular trend or a temporary change remains to be seen. In any even, electricity's share of total energy consumption increased dramatically in recent decades, growing at two and a half to three times the rate of total energy.[4]

As electrical output has grown, so has concern over the environmental issues involved in the generation of electricity. The main purpose of this chapter is to discuss the effects of conventional steam electric generation on air quality and their relation to coal use.

Electricity generation presently comes from three sources: hydroelectric, nuclear, and fossil fuel. Statistics are reported for electricity generated by internal combustion processes. In 1972, less than one-half of one percent of electricity was generated by these methods.[5] Even though net additions have been made, hydroelectric power has become relatively less important over time. In 1961, 19.2 percent of all power generated was hydroelectric, but this fell to 15.6 percent in 1972.[6] This relative decline has occurred for two main reasons. As dams are built and turbines installed, there are fewer suitable sites for hydroelectric generation. The Tennessee Valley Authority's work in the Southeast, for example, provided large hydroelectric power sources in the 1930s, but there have been few net additions since that time. Second, the growing concern over damming and flooding the country's remaining free-flowing river and stream resources has increased the resistance to new dams.

Nuclear power is considered to have great potential as an electricity source for the future. In particular, the development of breeder reactors, which can produce more nuclear fuel than they consume, is determined to be a major step. At present, nuclear power plants are similar to conventional steam electric plants in that both generate steam to drive a turbine which produces electricity. In 1972, nuclear steam generated 3.1 percent of total electric power.[7] If all planned additions to nuclear capacity were operational, the nuclear share of electricity generation would be only 5 percent by the end of 1975.[8]

There have been problems, however, in meeting installation schedules. Completion delays for large nuclear units kept almost two-thirds of planned additions off the line in 1973.[9] Though some set-backs are due to technical difficulties, environmental action groups have often caused interventions and licensing delays. The safety of operational nuclear materials has been questioned, not only because of possible accidents, but also because nuclear systems regularly discharge low levels of radioactivity. Nuclear wastes from these plants are highly radioactive, requiring special care in transporting and disposal. The disposal is really a storage operation, because the long half-life of nuclear wastes requires that they be monitored indefinitely. There have been proposals to bury these wastes deep underground in order to minimize the danger of their radioactivity, but at present most nuclear wastes are held above ground.

The major source of electric generation is through conventional steam turbines powered by fossil fuel. In 1972, 96.3 percent of steam electric generation was from conventional steam with only 3.7 percent from nuclear plants.[10] Thus the three fossil fuels—coal, fuel oil, and gas—are the major fuel inputs in electric generation.

WATER-QUALITY ISSUES

The environmental issues in steam electric generation are concerned with air and water quality. Damage to water quality stems mainly from thermal discharges because it is impossible, for theoretical and practical reasons, to convert all the energy in a fuel directly into electricity.[11] The heat rate, that is, the number of Btus necessary to produce one kilowatt hour of electricity, theoretically is 3,413. But the average in most electric generating plants at present is 10,000 Btus or more, even though rates have decreased greatly over the long run. The excess heat is released into the environment. In steam electric plants powered by fossil fuels, 15 percent of the heat is released into the air, and 85 percent remains in the condenser. The excess heat is removed from the condenser by running water around it.

Thermal discharges from nuclear plants are even larger than those from fossil-fuel plants because plant efficiency is around 30 percent less and the system is closed for protection from radiation. Thus 95 percent of waste heat is released in coolant water and only 5 percent directly into the air. At any rate of generation, the light-water reactors either planned or presently in use require about 50 percent more cooling water for the condenser than modern fossil-fuel plants.[12] In fact, modern nuclear reactors require 25 percent more cooling water than older less efficient fossil-fuel plants.[13]

The environmental effects of thermal discharges are not well known or quantified. The tolerance of fish and the reaction of algae have been studied, but questions remain concerning "the transport and behavior of heated water in streams and reservoirs, the biological effects of thermal discharges on aquatic organisms, and improvements in cooling systems."[14] The heat rate for coal is less than those for oil and natural gas. In 1970, the heat rate of coal was approximately 9 percent less than the rate for oil and 5 percent less than natural gas.[15] This average includes the higher heat rates of the older, less efficient coal-fired plants. In 1967, for example, the average value for the ratio of waste heat released to the electricity produced was 2.0, but the rate was only 1.5 in the best coal-fired plants.

AIR-QUALITY IMPACTS AND PROPOSALS

To the best of our knowledge, the environmental issue of thermal discharges has not materially affected the choice of fossil fuel.[16] But such selection has been affected by air-quality issues. Electric utilities form a major sector of stationary combustion sources in our economy. The three main air pollutants emitted in electric generation are sulfur oxides, nitrogen oxides, and particulates. Table 2.1 lists the major contributors of air pollutants by source in 1970. As we can see, sulfur oxides form the major pollutant from electric generation, and electric utilities are the main source of sulfur oxides. Electric generation is the source of over half the emitted sulfur, one-fifth of nitrogen oxides, and almost one-sixth of particulates.

In 1973, coal was the source of approximately 56 percent of conventional steam electric generation with the share of oil and gas at 21 and 23 percent, respectively.[17] The physical and chemical properties of natural resource deposits, even for the same fuel, may vary according to location, but coal generally produces more of the three main utility pollutants than either oil or gas.[18] The majority of coal deposits currently mined and used by utilities contains higher levels of sulfur than most of the oils and natural gas. They generate more particulates and emit more nitrogen oxides than oil and gas. Therefore, any discussion of pollution from electric utilities often centers around coal.

Of the three emissions, particulates are the object of the most advanced technological control systems, probably since they have caused the greatest concern historically. The effects of particulate pollution, namely, smoke, ashes, and soot, are more noticeable than gas emissions. Particulates create damages in

TABLE 2.1

Contribution to Air Pollution by Major Sources, 1970
(percent)

	Carbon monoxide	Particulate	Sulfur oxides	Hydrocarbons	Nitrogen oxides
Transportation	75.5	2.6	2.9	56.2	51.5
Motor vehicles	65.7	1.6	0.9	48.1	40.1
gasoline	65.2	1.2	0.6	47.8	34.4
diesel	0.5	0.4	0.3	0.3	5.7
Aircraft	2.0	0.2	0.3	1.2	1.6
Railroads	0.1	0.2	0.4	0.3	0.6
Vessels	1.2	0.2	0.9	0.9	0.7
Nonhighway motor fuels	6.5	0.4	0.5	5.8	8.5
Stationary com-bustion sources	0.5	26.8	78.2	1.8	44.1
Utilities	0.1	14.5	57.2	0.4	20.7
Other industry	0.0	8.4	14.4	0.8	20.0
Residential and commercial	0.4	3.9	6.4	0.6	3.5
Industrial processes	7.8	52.0	17.8	15.8	0.9
Solid waste disposal	4.9	5.6	0.3	5.8	1.7
Other	11.4	13.3	0.8	20.6	1.8
Forest fires	1.7	3.3	–	0.6	0.4
Other fires	0.1	0.2	–	0.2	0.0
Coal refuse fires	0.2	0.4	–	0.2	0.1
Agricultural burning	9.4	9.4	–	7.9	1.2
Organic solvent	–	–	–	8.8	–
Gasoline marketing	–	–	–	2.9	–

Note: Numbers may not add due to rounding by source.

Source: United States Environmental Protection Agency, *Nationwide Air Pollutant Emission Trends 1940-1970* (Research Triangle Park, North Carolina: Environmental Protection Agency, 1973), p. 44.

21

TABLE 2.2

Control of Fly Ash at TVA Coal-Fired Power Plants

Year	Removal Efficiency, percent	Approximate System Generation[a]
1950	0	20
1955	66	44
1960	71	65
1965[b]	57	82
1970	87	100
1977[c]	99	—

[a] In billion kilowatt hours.
[b] Represents transitional switch to electrostatic precipitators.
[c] Projected.
Source: Tennessee Valley Authority, *A Quality Environment in the Tennessee Valley* (Knoxville, Tennessee: Tennessee Valley Authority, 1974), p. 16, and *TVA Today 1974* (Knoxville, Tennessee: Tennessee Valley Authority, 1974), p. 28.

two areas. First the additional soot and dirt necessitate increased maintenance in the general air shed around the generating plant. Second, human health is affected by breathing the particles, though respiratory problems may result more from the particulate "transporting" dissolved chemicals than from the particles themselves. The interaction between sulfur oxides and particulates is one example.

The experience of TVA serves as a good example of particulate control. Until 1950, no fly ash was removed by control devices but was emitted from the stacks at TVA coal-fired plants. Mechanical ash collectors were the first devices installed to help control particulates and eventually these were replaced by electrostatic precipitators. As shown in Table 2.2, the resulting decrease in particulate emissions since 1950 has been dramatic. Use of low sulfur coals could interfere with precipitator operations achieving the projected 99 percent efficiency rate, at least with the collection system installed in the mid-1960s.[19] However, it also appears that a different collection system is possible, allowing much higher gas temperatures to overcome any particle resistivity of low sulfur coal.[20] The technology in particulate control has generally been effective, but TVA is conducting a program to improve it.

The effect of nitrogen oxide as a pollutant is not certain at present, and there is little public pressure to control its emission. A similar situation existed 10 years ago, when the emission of sulfur oxide was recognized but its potential damage was not known. Consequently, no firm public policy toward control was established. As reported by Gordon, the Environmental Protection Agency

(EPA) criteria report on nitrogen emissions suggests that, except for the effect on visibility, the damages are rather uncertain.[21]

At the highest concentration levels, EPA indicates there may be "some" harm to human health and vegetation. Nitrogen oxides are formed when the nitrogen compounds in fuel are freed during combustion and also when nitrogen in the atmosphere reacts with oxygen released during combustion. The main efforts at control are modification of the process of combustion through double combustion systems or recirculation of the stack gas. As noted in the recent voluminous EPA report summarizing technical studies on fuel energy systems, "The impact of nitrogen oxide emissions in the future certainly will be lessened by the use of combustion modifications, but the exact extent is difficult to forecast now."[22] The control of nitrogen oxide emissions is less successful when oil is the fuel input for utilities than when gas is used, but the technical reasons for that are not entirely clear.[23] Coal, with a more involved combustion process, might have the greatest control problem through combustion modification. Research by Exxon for the EPA showed few control problems with coal in utility boiler tests.[24] At present it is difficult to forecast the impact that nitrogen oxide, as a pollutant, will have on the use of coal as an electric utility fuel.

Potentially a pollutant could interfere with coal use if public pressure to reduce its emission was well established and if fully operational control technology was not widely available or in use. Sulfur oxides, which are emitted in the combustion of sulfur-containing fuels fulfill these conditions. Though all fossil fuels have some degree of sulfur content, sulfur compounds pose the greatest difficulty in coal. For example, natural gas as fuel has almost no sulfur content, not because it is inherently low in sulfur, but because the sulfur is removed before initial delivery to avoid pipeline corrosion. The technology for this removal is well developed and a normal process in natural gas operations. Oil use presents more complicated problems. Residual oils, the main oil used by utilities, can actually be higher in sulfur than the crude oils from which they were refined.[25] Sulfur levels are generally influenced by different oil blendings and by the availability of treatment plants for sulfur removal.

Efforts to reduce the sulfur level of coal before shipment have not had much success. The sulfur is found in coal in two forms, organic and pyrite. Pyrite sulfur, which varies among coal fields from 4 to 60 percent of total sulfur, can be significantly reduced by mechanical cleaning and crushing.[26] At present, around 50 percent of all coal produced is cleaned, but, since organic compounds are unaffected by cleaning, these operations usually remove less than half of the sulfur.[27] The only possibility of eliminating sulfur from coal appears to be through the development of a method of producing synthetic fuels from coal, particularly synthetic gas.[28]

In the metallurgical coal market, cleaning is a routine step in preparing high-grade coal for coking and related uses. Other cleaning operations are undertaken explicitly to reduce the sulfur levels in coal for the electric utility market.

In Illinois, 94 percent of all coal produced in 1972 was for electric utilities.[29] The average sulfur content, as delivered, was 3.4 percent, a fairly high level.[30] Yet almost all of this coal had been first mechanically cleaned at the mines.[31] Thus, the removal of pyrite sulfur alone, while it improves the quality of coal somewhat, cannot reduce the sulfur content to a sufficient degree.

The effects of sulfur oxides as a pollutant are not completely understood. They can create a mild acidic solution in water which is corrosive to structures and harmful to vegetation. High levels of sulfur oxides and concentrated exposure periods can dangerously aggravate respiratory ailments. Also the health hazard of particulates is substantially increased if sulfur oxides are also present.

AIR POLLUTION AS AN EXTERNAL COST OF PRODUCTION

Air pollution is a matter of concern to economists since it represents an external cost of production not accounted for in private market activity.[32] The

FIGURE 2.1

Private Production Costs and the Social Costs of Externalities

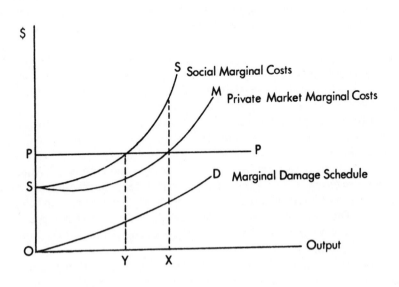

FIGURE 2.2

Benefits and Costs of Abatement

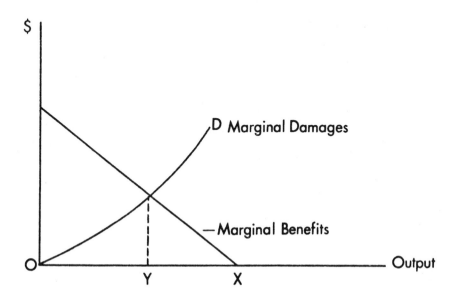

use of air for free disposal of waste emissions can result in damages which are not reflected in production costs and, therefore, are not stated in the value of resources used in production. Suppose a firm emits a single pollutant that causes damages for which the firm does not compensate as in Figure 2.1. The demand curve for our competitive firm is PP, which corresponds to the value consumers have placed on the product. The private marginal cost schedule of the firm is SM, which does not include any damage costs. Assuming that marginal damages increase with output, the social cost of production is SS, total value of resources used in the production process. It is the private marginal cost schedule of the firm plus the marginal damage schedule, OD.

The private firm, attempting to maximize profits, would produce at X, where private marginal cost equals marginal revenue. At this level of production, however, social marginal costs are larger than marginal benefits to society. The point at which the total resource marginal cost equals marginal benefits is at a production level of Y. The problem for public policy is to reduce the firm's production from X to Y.

Alternatively, consider Figure 2.2. The marginal benefit to the firm of increased output is the difference between the marginal cost of producing additional output and its marginal revenue in Figure 2.1. If there were no external costs, these marginal benefits would be the same for society as well as the private firm. The firm, considering only its private marginal costs, would produce to the point where marginal benefits were no longer positive, at output X. But the marginal damages from the emission are OD as before. As output is increased from zero, only at their intersection is the marginal increase in benefits to society equal to the total marginal costs of production. This is, of course, at output level Y in Figure 2.1.

In order to establish optimal levels of emissions as a matter of policy, we would need correct measures of both the value of damages and the costs of abatement. At alternative levels of emissions, the damages incurred by society could then be related to the cost of abatement measures, and we would employ resources to control pollution up to the point where their value was just equal to the reduction in damages. Attempting to measure the relevant variables is difficult. The physical damage functions are often concerned with public goods, where it is difficult to place a market value on reduced health hazards, increased visibility, and the aesthetic improvement of the environment. As Gordon puts it, "Expressing the ideal is virtually equivalent to showing that it is unattainable. Neither the damages nor their controls are sufficiently understood to permit even the grossest approximation of a measure of optimal policy."[33]

Some research has been done relating the physical damage from sulfur dioxides and particulates to property values.[34] As expected there is an inverse relationship between property values and levels of pollution, but the resulting estimated coefficients are highly sensitive to the model's specification.[35] The wide range of variables influencing property values and the inadequacy of tax assessment figures as a measure of property value make these estimates even more tenuous.[36]

One of the most widely cited studies of damage estimates is by Barrett and Waddell.[37] They found that out of a national total of $16 billion in air pollution damages in 1968, sulfur emissions were responsible for $8.3 billion— with $2.8 billion to the residential sector, $2.2 billion to materials, and $3.3 billion to health. Of the remaining total damages $5.9 billion were attributed to particulates, with $2.4 billion to the residential sector and $2.8 billion to health.

If we can distribute damages in proportion to the physical tons of particulates and sulfur emissions, the 33.2 million tons of emitted sulfur produced damages of $250 per ton of sulfur oxide.[38] Given our earlier ratio of 1.9 between sulfur oxides and sulfur emissions, this represents approximately $500 per ton of emitted sulfur in 1968 dollars. If these damages have increased proportionately to the gross national product (GNP) price deflator since that time, the damage figure in 1972 would be $600 per ton.

Attempting to achieve "efficient" air quality through standards administered by a regulatory public agency is the most common approach to control. The agency has a rather complex task in setting its standards. The physical damage function for pollutants must be determined, the level of feasible control technology must be considered, and the air-quality goals must be set in conjunction with these facts. Since each pollutant is considered separately in setting regulations, the data requirements necessary to make a correct judgment are formidable. Usually the agency has a technical staff to monitor emissions, undertake research, etc., and funds private or academic institutions to undertake research projects. The federal government at present determines basic standards through the EPA. The states must submit their specific programs to meet these standards with the EPA, which has legal power to determine its own plan if it finds the state action inadequate.

The determination of air-quality goals is made more complex by the fact that each pollutant is subject to two standards. A primary, or ambient, standard is designed to provide protection from any possible health hazards. A secondary standard is meant to eliminate "other" losses, such as aesthetic or vegetation damage. The process becomes even more difficult since these two standards for the same pollutant can vary between different emission sources. Electric utilities, for example, are subject to different particulate requirements than those for residential use. Emission standards are usually specified as allowable average emission concentrations per million Btu. Since several different measures are available, it is difficult to convert emission standards to a standard energy basis.[39] Consequently, the operational setting of standards can be controversial.

Given the present lack of knowledge concerning physical damage functions for each pollutant, the level at which standards are set is unclear. The dramatic effects of pollution, such as the deaths from concentrated emissions in the stagnant air shed of London in 1952, are often cited. Yet, if these cases were our only concern, standards to regulate each emission for all periods of operation might be unnecessary. We could simply employ an off-on strategy of standards when air conditions and pollutants built to critical levels, including, for example, production slowdowns as well as the use of less polluting fuels. A counter argument is that we actually are not adequately realizing or measuring the complicated long-term health effects of steady emission exposure.

If public agencies become biased in the administration of air-quality standards, they may attempt to avoid even relatively minor inconveniences. Policy can become inefficient if it attempts to be too rigorous. As shown in Figure 2.2, it is impractical to prescribe levels to the left of output Y. It has been reported that the approach to primary emission standards has focused on the worst possible effects under the worst possible air shed conditions, and that, at least for nitrogen oxides, they are based on discredited research.[40]

As we have seen, sulfur oxides pose more difficulty for the electric utilities than other emissions. One reason is that there is not a direct relationship between ambient air standards and specific sources, such as steam electric generating plants; it is largely a matter of judgment. For example, the EPA's own guidelines for applying emission standards to new plants are based on what are believed to be practical emission levels under new technology.[41] An unpublished study done for EPA shows that many regions could meet the ambient air standards for sulfur oxides under much less stringent source controls than have been proposed.[42] A second question concerns the point at which emissions would be measured. In an effort to disperse emissions over a wide area and to reduce ground level concentrations of the furnace gases and ash particles, utilities have installed tall smokestacks at power plants. These stacks usually range from 500 feet to 1,000 feet, which is about twice the height of the Washington Monument.[43] In general, the electric utility industry has been a strong advocate of this control measure, and satisfactory results for ground level emissions were claimed.[44] However, the EPA standards, as amended in 1971, measure emissions from the point of their entry into the atmosphere, at the top of the stacks. As a result, these tall stacks are not in compliance and the question remains concerning the best point for measuring the accumulation of pollutants.

In the 1960s, the development of stack-gas cleaning methods for removing sulfur oxides after combustion appeared to be the most promising control technology. It was felt that the time and money needed for developing such systems were much more acceptable than those for developing synthetic coal products. Since there was very little development of the low-sulfur Western reserves at this time, emphasis was placed on the development of scrubbing systems for meeting emerging sulfur standards. It was thought that relatively inexpensive scrubbing systems would be operational by the mid-1970s. Since that time, skepticism over the reliability of such systems has increased markedly. Though some utilities, such as the ones in Louisville, Kentucky, consider operational scrubbers as a feasible solution to sulfur emissions, it is probably fair to say that, in general, the electric utility industry views the scrubbing concept as a colossal fiasco.[45]

Before considering the quantitative operating evidence on these systems, there are two additional considerations. The scrubbing systems can be classified into two general types, throwaway systems and by-product recovery systems.[46] In the throwaway systems, sulfur removed from the stack gas must be disposed of as a waste. By-product systems attempt to recover and transform the captured sulfur into a useful chemical product, such as sulfuric acid or elemental sulfur. There are several different types of scrubbing systems under development which are classified according to the main chemical agents used, such as wet limestone. Since the technology of the throwaway system at present is much more advanced than that of the by-product systems, the first large-scale installation attempts would undoubtedly be of this type.

Waste from a throwaway system is a sludge material, and its disposal is a matter of environmental concern that is not usually considered in the literature favoring scrubbing. The importance of this issue may well be that the technology of throwaway systems is generally site specific.

A second concern is the ability of a stack-gas cleaning industry to supply major demand for systems in a reasonable time frame. Because it is uncertain at present which system will emerge as the principle one, a large supplying industry has not been established. It is estimated that over $4 billion would be needed to develop a stack-gas cleaning industry.[47] Furthermore, there is concern about the time lag between production orders for systems and the date of their on-line availability. An interesting report for EPA indicates a lead time of approximately three years.[48] Concern has been expressed over possible short-run bottlenecks in skilled personnel to operate and maintain these systems if their use increases greatly.[49] The implication is that, even if reliable systems were developed overnight, society would not find them instantly available.

The main issue concerning scrubbing systems has centered on their performance and reliability. As late as 1970, the National Coal Association stated that the solution to sulfur emissions through scrubbing technology was imminent.[50] Consider the two cases cited by the National Coal Association as near success. Union Electric and Combustion Engineering jointly spent over $3 million to develop a pilot project near St. Louis, but due to continual plugging in the boiler, the system has now been abandoned.[51] The pilot system at Kansas Power and Light has received extensive modifications over the past four years and has not yet achieved successful operation. The experience of other utilities has not been encouraging, though some still feel that scrubbing is feasible.

The operating data now available are too limited, making it difficult to assess the situation to find positive conclusions and to make generalizations for the entire industry. General observations on scrubbing systems can only be considered in light of specific attempts to apply them. The chief support for scrubbing technology is contained in a government report of an interagency committee considering scrubbing systems.[52] Their optimistic conclusions, though not unanimous—in particular, the commission member from the FPC dissented from the conclusion—were based on two main observations. The first was based on the findings of a team studying sulfur-emission-control technology in Japan in 1972 under the committee's auspices. In addition to its primary study of oil controls, the study team examined a coal-fired plant that had successfully scrubbed flue gases for four months. In 1973, it was still the only existing coal-fired plant able to accomplish this on a normal operating basis.[53] The study team's observations proved that operational scrubbing was possible, but that there were several qualifications to its applicability to this country.[54] The boiler size was small compared to the average U.S. system. Higher input concentrations of both ash and sulfur dioxide occur in this country and the variance in boiler loads is higher.

Therefore, the same system is not necessarily suitable in this country, even though the technology at the Japanese plant is of U.S. origin and design.

The second observation was the "promising" work of Commonwealth Edison. Their research experience with scrubbers has not made the company optimistic over its potential. The dual scrubbers installed at their Chicago plant have had a simultaneous availability of only 8.1 percent of the time.[55] Component failures of such systems could be lessened through redesign, but the problems of plugging and scaling still remain.

It is difficult, at best, to determine when systems will be proved entirely reliable. Consider the time involved for processes such as double alkali or Wellman-Lord systems. The pilot plants erected to begin tests for double alkali began in 1961.[56] Wellman-Lord advertised a system of "90-percent efficiency" in 1970.[57] Yet these two systems are still not on the market and the latest estimate is that they will be ready for market inspection early in 1977.[58]

Given the problems which have occurred in the past, this projection could be optimistic and, in fact, any on-line scrubbing on a broad scale does not appear likely at least until the early 1980s.[59] As the technical summary report from EPA indicates, "As operated today, power plants have no control over SO_2 emissions other than by selection of fuels."[60] While the potential technology exists, the short-term outlook for these systems does not appear promising. In this study, we do not explicitly consider scrubbing as a method of reducing sulfur emissions in the model.

An alternative to setting standards to achieve air quality is to levy a tax which would be equal to the damages of the emissions. Such a tax would equate private and social costs of production, and the external effects would be internalized as costs to the polluting firms. The levying of a tax on sulfur emissions could provide an economic incentive to reduce emissions. The agency responsible for administering air-quality programs theoretically would still need to identify the physical damage function to properly set the rate. Yet unlike standards where the agency must consider available control techniques, a sulfur-emissions tax would provide the incentive for the utility or firm to choose the most efficient control technology. In particular, it would be financially advantageous for polluters to invest in research and development to lower their emissions in the future. A sulfur-emissions tax would impose a user cost that would vary with the sulfur content of the fuels used and with control technology. As a result, it could affect either aspect of the process. First, it could stimulate the development of control technology, particularly scrubbing systems. Second, it could switch to lower sulfur fuels in order to avoid the tax. Such a taxing scheme is seen as a viable alternative to set standards, since electric utilities can challenge or evade most present air-quality standards through lengthy litigation.[61]

The proposed tax rates generally run from 5 to 15 cents per pound of emitted sulfur, with a rate of 10 cents being the most widely considered. An alternative proposal has been to put the sulfur tax directly on the fuel inputs. In a static model, the effects of taxing either the sulfur in the fuel input before combustion or the inherent sulfur content contained in the fuel as measured by the stack emissions are essentially the same. Yet over time the direct fuel input proposal would only tend to provide the incentive to use low sulfur fuel, with any impetus to develop emission control technology being reduced. In Chapter 7, we will consider the effects of a sulfur-emission tax and will comment on the idea of simply prohibiting fuel inputs whose sulfur contents are greater than some "critical" level.

NOTES

1. *Electrical World,* "Twenty-Third Annual Electrical Industry Forecast," 1972, p. 64.

2. Harold Barnett, *Energy, Resources and Growth* (St. Louis: Washington University, Department of Economics, 1973), p. 22.

3. Ibid., pp. 25-26.

4. Oran L. Culbertson, *The Consumption of Electricity in the United States* (Oak Ridge, Tennessee: Oak Ridge National Laboratories, June 1971), p. 7.

5. Edison Electric Institute, *Statistical Yearbook* (New York: Edison Electric Institute, 1973), p. 20.

6. Ibid., Table 125, p. 20.

7. Ibid.

8. For a general discussion of nuclear power, see the report of the Joint Committee on Atomic Energy, *Nuclear Power and Related Energy Problems: 1968 Through 1970* (Washington, D.C.: Government Printing Office, 1971). The data given are from p. 377.

9. *Electrical World,* op. cit.

10. Edison Electric Institute, op. cit., p. 22.

11. Except as noted, our discussion is based on Sylvian Denis, *Some Aspects of the Environment and Electric Power Generation* (Santa Monica, California: Rand Corporation, 1972), pp. 13-17.

12. Federal Power Commission, *Problems in Disposal of Waste Heat from Steam Electric Plants* (Washington, D.C.: Government Printing Office, 1969), p. v.

13. Ibid., p. vii.

14. Ibid., p. viii.

15. National Coal Association, *Steam Electric Plant Factors, 1972* (Washington, D.C.: National Coal Association, 1973), p. 102.

16. However, the need for coolant water can affect the selection of power-plant construction sites. See Executive Office of the President, Office of Science and Technology, Energy Policy Staff, *Considerations Affecting Steam Power Plant Site Selection* (Washington, D.C.: Government Printing Office, 1970).

17. Federal Power Commission, *Monthly Report of Cost and Quality of Fuels at Steam Electric Generating Plants* (Washington, D.C.: Government Printing Office, December 1973), p. 1.

18. See the respective emission factors as discussed in United States Environmental Protection Agency, *Compilation of Air Pollutant Emission Factors* (Research Triangle Park, North Carolina: Environmental Protection Agency, 1973).

19. Tennessee Valley Authority, *TVA Today 1974* (Knoxville, Tennessee: Tennessee Valley Authority, 1974), p. 28.

20. Ibid.

21. Richard L. Gordon, *U.S. Coal and the Electric Power Industry* (Baltimore: Johns Hopkins University Press, 1975). Chapter 6 provides a discussion of major utility pollutants.

22. United States Environmental Protection Agency, Office of Research and Development, *Environmental Considerations in Future Energy Growth*, Vol. I, *Fuel-Energy Systems* (Washington, D.C.: Government Printing Office, 1973), p. 388.

23. Ibid., p. 388.

24. Ibid.

25. For sulfur considerations in oil and different import blendings, see Richard Gordon, "Environmental Impacts in Energy Production and Use," "Energy Supply Project" (unpublished manuscript on file, Resources for the Future, Washington, D.C., 1973), pp. 115-16.

26. Congressional Research Service, *Factors Affecting the Use of Coal in Present and Future Energy Markets* (Washington, D.C.: Government Printing Office, 1973), pp. 31-32.

27. Ibid., p. 32.

28. A discussion of synthetic processes occurs in Chapter 5 of Richard Gordon, *U.S. Coal and the Electric Power Industry*, op. cit. A summary discussion of these processes occurs in Congressional Research Service, op. cit., pp. 133-34.

29. Ibid., p. 83.

30. Ibid.

31. Ibid., p. 63.

32. Our discussion in this section is drawn from Robert Bish, *The Public Economy of Metropolitan Areas,* Markham Public Finance Series (Chicago: Markham Publishing Company, 1971), pp. 122-28. We follow his general diagrammatic presentation, particularly his Figure 2.

33. Gordon, "Environmental Impacts of Energy Production and Use," op. cit., p. 11.

34. See, for example, R. Ridker, *The Economic Costs of Air Pollution* (New York: Praeger, 1967), and R. Ridker and J. Henning, "The Determinants of Residential Property Values with Specific Reference to Air Pollution," *Review of Economics and Statistics* 49 (1967): 246-57.

35. James Griffin, "Recent Sulfur Tax Proposals: An Econometric Evaluation of Welfare Gains," in *Energy: Demand, Conservation, & Institutional Problems,* ed. Michael S. Macrakis (Cambridge: Massachusetts Institute of Technology Press, 1974), p. 237.

36. As discussed by L. Lave, "Air Pollution Damages: Some Difficulties in Estimating the Value of Abatement," in A. Kneese et al. (eds.), *Environmental Quality Analysis* (Baltimore: Johns Hopkins Press, 1972), especially pp. 214-15.

37. The following figures are from their report, L. Barrett and T. Waddell, *Costs of Air Pollution Damages: A Status Report* (Research Triangle Park, North Carolina: Environmental Protection Agency, 1973), especially pp. 59-62.

38. For a critical review of their methodology, see Gordon, "Environmental Impacts of Energy Production and Use," op. cit., pp. 19-23. The health estimates, for example, are based on simply doubling a previous study which considered cutting pollution in half.

39. Gordon, *U.S. Coal and the Electric Power Industry,* op. cit., Chapter 6, p. 126.

40. Ibid., p. 3.

41. Ibid.

42. Gordon discusses this report in "Environmental Impacts in Eneergy Production and Use," op. cit., p. 25.

43. TVA, op. cit., p. 15.

44. Phillip Sporn (ed.), *The Tall Stack for Air Pollution Control on Large Fossil Fuel Power Plants* (Washington, D.C.: National Coal Policy Conference, Inc., 1967).

45. See the advertisement, for example, of the American Electric Power System in the Washington *Post*, August 13, 1974.

46. See Congressional Research Service, op. cit., pp. 32-33.

47. Ibid., p. 34.

48. Hittman Associates, Inc., *Assessment of 502 Control Alternatives and Implementation Patterns for the Electric Utility Industry* (Washington, D.C.: Environmental Protection Agency, 1973), p. 21.

49. Congressional Research Service, op. cit., p. 34.

50. National Coal Association, *Bituminous Coal Facts 1970* (Washington, D.C.: National Coal Association, 1971), pp. 22-24.

51. For an account of utilities' attempts to perfect scrubbing technology to date, see United States Environmental Protection Agency, Office of Research and Development, op. cit., p. 407, which is the basis for our discussion.

52. Sulfur Oxide Control Technology Assessment Panel, *Final Report on Projected Utilization of Stack Gas Cleaning Systems by Steam Electric Plants* (Washington, D.C.: Government Printing Office, 1973).

53. As reported in E. Erikson et al., *The Energy Question, Volume 2* (Toronto: University of Toronto Press, 1974), p. 60.

54. See Gordon, "Environmental Impacts of Energy Production Use," op. cit., pp. 118-19.

55. United States Environmental Protection Agency, Office of Research and Development, op. cit., p. 407.

56. A general description of the assortment of scrubbing technologies is contained in National Air Pollution Control Administration, *Control Technique for Sulfur Oxide Pollutants* (Washington, D.C.: Government Printing Office, 1969), pp. 48-57, particularly.

57. National Coal Association, op. cit., p. 22.

58. Hittman Associates, op. cit., p. 15.

59. Ibid. B-63 mentions that the Wellman-Lord 10-month period ending December 1975, reflects, for example, "a somewhat optimistic timetable estimate provided by E.P.A."

60. United States Environmental Protection Agency, Office of Research and Development, op. cit., p. 385.

61. The Council on Environmental Quality has been reported as feeling that the tax could be preferable to a standard for these reasons. See Congressional Research Service, op. cit., p. 34.

3

A SHORT-RUN
REGIONAL MODEL OF
STEAM ELECTRIC COAL

In this chapter, we examine the factors affecting the degree of regional dependence on coal for electric generation and the relationship between these factors and the regional distribution of total U.S. coal production. We then discuss a spatial linear programming model of the steam electric coal market, which will be used in future chapters to consider the regional implications of those public policies that are concerned with the utilization and production of coal.

REGIONAL FACTORS AND THE COAL INDUSTRY

Coal resources, like most natural resources, are nonrenewable stocks with fixed locations. The study of a resource that has a spatial dimension fixed by the nature of its available deposits must consider the regional implications of the analysis. Because of the impact of coal extraction on land use, these regional considerations are relatively more important for coal resources than for other fossil fuels. The Bureau of Mines has classified the coal fields of the United States into a set of 23 coal-supply districts,[1] as seen in Figure 3.1. The classification is based on the similarity of conditions for coal production in each district. This classification scheme is a traditional one, originally outlined in the Bituminous Coal Act of 1937. This is due partly to the fact that researchers have found the data published by the Bureau of Mines for this regional delineation to be the most detailed. For our purposes, this classification has been surprisingly good. Districts 1 and 2, for example, are both in Pennsylvania. The coal fields of District 2 supply the industrial centers of southern Pennsylvania, while District 1 is the primary source of shipments to electric utilities in the northeast.

FIGURE 3.1

Fuel Supply Districts: Coal

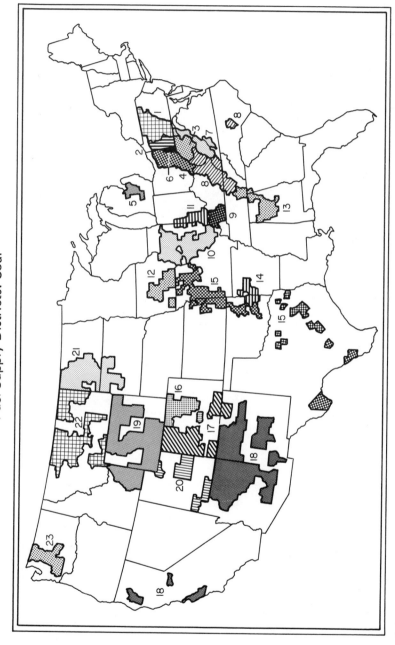

TABLE 3.1

Coal Shipments to Electric Utilities by District of Origin

District of Origin	1973[a]	Percent by Mining Method	
		Surface	Underground
1 Eastern Pennsylvania	34,362	50	50
2 Western Pennsylvania	9,403	19	81
3 and 6 northern West Virginia and panhandle	34,818	23	77
4 Ohio	38,926	68	32
7 Southern, numbered 1	789	12	88
8 Southern, numbered 2	65,747	35	65
9 Western Kentucky	52,894	65	35
10 Illinois	49,705	52	48
11 Indiana	20,454	94	6
12 Iowa	618	59	41
13 Southeastern	11,628	62	38
14 Arkansas-Oklahoma	—	—	—
15 Southwestern	12,665	100	0
16 Northern Colorado	492	0	100
17 Southern Colorado	2,974	43	56
18 New Mexico	11,008	100	0
19 Wyoming	14,113	96	4
20 Utah	1,903	1	99
21 North and South Dakota	6,098	100	0
22 and 23 Montana and Washington	13,567	99	1
Total, United States	382,164		

[a]In 1,000-ton units.

Source: National Coal Association, *Bituminous Coal Data, 1973* (Washington, D.C.: National Coal Association, 1974), p. 85.

Similarly, Districts 8 and 9 split Kentucky into eastern and western coal fields. The western fields of Kentucky are a major source of surface-mined, high-sulfur coal, while the eastern fields are an important area for lower-sulfur, underground coal.

These coal fields are not utilized equally. From Table 3.1 we see that the major areas of production for electric utilities are centered in Appalachia and the Midwest. We can identify several factors that contribute to this. Suppose we had an economy on a flat plane, with demand for the natural resource distributed evenly, production costs equal at all points, homogeneous product, and zero transport costs. Then we could not predict any general tendencies for the location of production. Indeed, unless there was some type of variable external costs or similar consideration, the location of the natural resource extraction would not be an economic issue.

As we begin to relax our simplified assumptions, production can be predicted. Assume electric utilities attempt to minimize the costs of their fuel purchases, and let transport costs be positively related to distance. Since delivered price includes extraction costs plus transport costs, market areas for each supply district would be delineated. As long as a district could supply utilities at a lower delivered price than its competitors it would include these customers in its market area. Since the coal deposits are scattered in a nonuniform pattern, the size of market areas and the amount of production would be different. Transport costs are a major component of the delivered price of coal, as much as 42 percent in some areas.[2]

The demand by electric utilities for steam electric coal is not distributed evenly across the continental United States. Table 3.2 clearly shows that the Eastern and Midwestern markets make the heaviest use of coal for conventional steam electric generation. The fossil fuels, with relatively high transport costs, can be more competitive against alternative fuels in markets near their production.

The main production centers of natural gas are located in the southwestern Central states, California, and Arizona. As a result, natural gas has been the fuel favored by utilities in these areas. The regulatory structure of the natural gas industry has also contributed to this pattern.[3] Since intrastate sales are not subject to the federal price regulations, utilities in these production areas have been able to bid higher prices for the natural gas production.

The use of oil over coal has depended upon the availability of imports. In 1964, coal generated 70 percent of electricity in the Northeast.[4] Since the 1966 exemption of East Coast residual oils from oil import quotas, coal's share of the market has decreased. In 1972, coal supplied only 19 percent of generated electricity in this area.[5] An interesting point concerning this increase in oil usage on the East Coast has been brought out in a survey reported by Gordon.[6] The major impetus to increased oil use during this period has been the lowering of its price in relation to coal, not environmental pressure from sulfur emissions. Most utilities would have adapted to oil-fired capacity because of price considerations alone, regardless of emission considerations.

When we look at areas where natural gas and oil do offer these advantages, we find that coal is clearly the dominant fuel. When we recognize the importance of the interaction between transport costs and the geographical distribution of coal demand, we can see why coal production is centered in the Appalachian and Midwestern districts. These districts, close to major markets, enjoy a delivered cost advantage over more distant coal producers in the West.

TABLE 3.2

Type of Fuel Used in Conventional Steam Generation Electric Plants

Region and State	Percent of Total Btu		
	Coal	Oil	Gas
New England			
Connecticut	21	79	—
Maine	—	100	—
Massachusetts	2	95	3
New Hampshire	60	40	—
Rhode Island	—	84	16
Vermont	92	8	—
Total	13	85	2
Middle Atlantic			
New Jersey	26	65	9
New York State	28	60	12
New York City	7	77	16
New York State (except New York City)	67	28	5
Pennsylvania	83	16	1
Philadelphia	26	70	4
Pennsylvania (except Philadelphia)	99	1	—
Total	54	40	6
Northeast Central			
Illinois	79	6	15
Indiana	94	1	5
Michigan	83	8	9
Ohio	98	—	2
Wisconsin	88	2	10
Total	89	3	8
Northwest Central			
Iowa	58	—	42
Kansas	5	1	94
Minnesota	67	2	31
Missouri	80	—	20
Nebraska	32	1	67
North Dakota	100	—	—
South Dakota	57	12	31
Total	58	1	41
South Atlantic			
Delaware	72	20	8
District of Columbia	28	72	—
Florida	17	45	38

Region and State	Percent of Total Btu		
	Coal	Oil	Gas
Georgia	75	3	22
Maryland	62	38	—
North Carolina	98	—	2
South Carolina	72	5	23
Virginia	52	47	1
West Virginia	99	1	—
Total	62	23	15
Southeast Central			
Alabama	97	—	3
Kentucky	98	—	2
Mississippi	11	5	84
Tennessee	95	—	5
Total	89	1	10
Southwest Central			
Arkansas	—	16	84
Louisiana	—	1	99
Oklahoma	—	—	100
Texas	—	—	100
Total	—	1	99
Mountain			
Arizona	10	3	87
Colorado	55	2	43
Montana	89	—	11
Nevada	44	1	55
New Mexico	70	1	29
Utah	43	49	8
Wyoming	95	1	4
Total	56	4	40
Pacific			
California	—	26	74
Oregon	—	29	71
Washington	—	100	—
Total	—	26	74
United States, total	54	16	30
Coal competitive states, I[a]	68	18	14
Coal competitive states, II[b]	73	16	11

[a]Total United States, excluding southwest Central and Pacific regions.

[b]Total United States, excluding Florida, Arizona, Mississippi, southwest Central and Pacific regions.

Source: National Coal Association, *Steam Electric Plant Factors, 1973* (Washington, D.C.: National Coal Association, 1974).

TABLE 3.3

Known and Recoverable U.S. Coal Reserves, January 1, 1965
(million short tons)

Regional Reserves	Sulfur Content (by weight, dry basis)									Total	Percent Known to Total Reserves
	<0.7	0.8-1.0	1.1-1.5	1.6-2.0	2.1-2.5	2.6-3.0	3.1-3.5	3.6-4.0	>4.0		
Bituminous Coal											
Northern Appalachian											
Known	45	2,755	21,370	23,050	27,525	11,950	8,780	7,155	800	103,340	13
Recoverable	5	360	2,780	2,995	3,580	1,550	1,140	930	105	13,445	
Southern Appalachian											
Known	37,275	41,025	18,135	9,890	2,770	3,510	275	45	85	113,010	11
Recoverable	4,100	4,510	1,995	1,090	305	385	30	5	10	12,430	
Eastern Interior											
Known	195	1,355	10,075	9,135	7,440	32,300	67,065	86,755	35,525	249,845	25
Recoverable	50	340	2,520	2,285	1,860	8,075	16,765	21,690	8,890	62,465	
Western Interior											
Known	250	770	2,475	1,180	9,170	2,070	11,340	28,975	62,685	118,915	8
Recoverable	20	60	200	95	735	165	905	2,320	5,015	9,515	
Northern Rockies											
Known	6,275	6,815	205	395	400	175	40	25	590	14,920	11
Recoverable	690	750	20	45	45	20	5	0	65	1,640	
Southern Rockies											
Known	38,940	56,295	0	1,525	0	0	0	0	3,995	100,755	10
Recoverable	3,895	5,630	0	150	0	0	0	0	400	10,075	
West Coast											
Known	900	685	0	0	0	0	0	0	0	1,585	9
Recoverable	30	60	0	0	0	0	0	0	0	140	

Subbituminous

Northern Rockies											
Known	129,665	109,045	0	1,300	0	0	0	0	10	240,020	11
Recoverable	14,265	11,995	0	145	0	0	0	0	0	26,405	
Southern Rockies											
Known	52,055	16,910	150	0	0	0	0	0	0	69,115	10
Recoverable	5,205	1,690	15	0	0	0	0	0	0	6,910	
West Coast											
Known	3,780	585	0	0	0	0	0	0	0	4,365	9
Recoverable	340	50	0	0	0	0	0	0	0	390	
TOTAL BITUMINOUS AND SUBBITUMINOUS											
Known	269,380	216,240	52,410	46,475	47,305	50,005	87,500	122,955	103,690	1,015,960	12
Recoverable	28,650	25,445	7,530	6,805	6,525	10,195	18,845	24,945	14,475	143,415	
TOTAL ANTHRACITE											
Known	12,550	93	0	145	285	0	0	0	0	13,075	13
Recoverable	1,630	10	0	20	35	0	0	0	0	1,695	
TOTAL LIGNITE											
Known	344,620	61,385	41,165	0	0	465	0	0	0	447,635	11
Recoverable	37,905	6,750	4,530	0	0	50	0	0	0	49,235	

Total known reserves* . . . . $1,477 billion
Total recoverable reserves . . . . $194 billion

* At close to present prices.
Source: Mitre Corporation, *Survey of Coal Availability by Sulfur Content,* Environmental Protection Agency (Washington, D.C.: Government Printing Office, 1972), p. 22.

Other factors can also be important. Extraction costs are not the same for all methods of mining. Large, powerful capital equipment has been developed making surface mining cheaper than underground mining.[7] The growth of surface mining has been quite dramatic as a result. In 1950, 24 percent of production was from surface mines. By 1972, surface mines accounted for 49 percent of total production.[8] The relative importance of these two mining methods varies across supply districts, as seen in Table 3.1. Surface mining is predominant in Western states. With the exception of District 8, which is a center for underground mining, the two methods are used almost equally in the majority of Appalachian districts. Surface mining is more important in the Midwest than underground operations. Missouri, for example, has only surface production. Approximately three-fourths of total surface-mine production is delivered to steam electric plants, which is 61 percent of coal delivered to utilities.[9] Since delivered price equals extraction cost plus transport cost, underground production close to major markets can be less expensive for utility customers than surface production from more distant centers.

Furthermore, coal is not a homogeneous product. The main technical consideration in electric generation is the ability of coal to produce steam, measured in British Thermal Units (Btu). Underground production exposes the coal seam to less atmospheric moisture, surface water runoff, etc., than surface mining. As a result, the average heat content of underground coal is about 8 percent higher than that for surface-mined coal. This difference is greater in the lower-sulfur coals, generally from 12 to 14 percent more.[10] On a price-per-million-Btu basis, this helps offset the disadvantage of underground mining in extraction cost per ton.

The sulfur content of coal can differ from one region to another. As we have seen, sulfur is the chief chemical property of coal that is subject to air pollution regulations. Table 3.3 gives a general picture of how coal reserves differ in sulfur content. Most of the low-sulfur coal is found in the West. The Midwest finds itself with relatively small low-sulfur deposits. The Appalachian area, contrary to popular belief, does contain significant amount of low-sulfur reserves. The heating value of Western coals, however, is generally lower than that of other areas. Therefore, a larger volume of this coal is necessary to replace coal from other areas on a per-ton basis. The apparent advantage that Western coal has in its low-sulfur content is offset by the greater amount needed to produce the same amount of energy. We will see the effects of this in Chapters 5 and 6.

In order to analyze coal reserves, we must consider the importance of transport costs, the location of major markets and resources, the differences in mining methods, and the nonhomogeneous nature of the product.

Leon Moses has characterized models of regional interaction and location interdependencies as "two varieties: (1) highly abstract formulations that extend Walras's reasoning to spatial phenomena, (2) models that are more restrictive in their assumptions and less general in their intent but which lend themselves to empirical application."[11] An example of the latter type is the study by Henderson, a linear programming model of the coal industry.[12] Henderson's model is closely related to the classic transportation problem in linear programming. The essence of this problem is to consider the firm with m warehouses and n stores. The capacity of each warehouse is known, and a fixed demand is required by each store. The firm's problem is to determine what shipments will minimize transport costs subject to the m capacity and n demand conditions. This particular problem's structure has been modified for numerous applications, even those which are not involved with transportation.[13] Henderson's modification recognized the interaction between the fixed location of coal deposits and the spatial distribution of demand. Using 1947 data, the study found the shipment pattern that minimized the cost of providing coal to these regions. Coal was differentiated by type and location of extraction. Our study is an extension of this modeling effort.

A REGIONAL MODEL OF THE COAL INDUSTRY

In presenting the model, we will first specify the behavioral assumptions. The model's basic activity is the delivery and utilization in a demand region of coal which has been extracted and shipped from one of the coal-supply districts. The demand regions consist of the 48 continental states and the District of Columbia. Up to 10 coal activities from any one supply district result from differentiating the coal by the two extraction processes (surface and underground mining) and five levels of sulfur. The utilization of any coal in steam electric generation is dependent on its energy value, measured in Btus. These energy values can differ by extraction method and by sulfur levels. Thus, differences by method of extraction, sulfur content, and Btu value for each supply district were considered separately. A major characteristic of the model allows us to consider values in this way as opposed to assuming one or two "general" values.

As interregional deliveries are varied by our analysis, the utilization of coal is considered independent of such additional characteristics as ash content, ash-fusion temperature, and agglomerating qualities. This point is mentioned because these characteristics in some Western coals may interfere with the function of certain wet-bottom boilers currently in use in the East. Three main coil boilers are used by utilities in conventional steam electric generation, i.e., cyclone, slag tap, and dry or wet bottom[14] The wet-bottom boilers can be modified for these Western coals, however, and in 1977 both wet- and dry-bottom boilers represents only approximately 10 percent of those used in electric utilities.[15] We do not believe our limiting assumption here contains any significant bias for our results.[16]

Our model is a short-run analysis. In his study, Henderson considered that "a short-run period, with reference to coal, is defined as a period of time during which the extraction functions and the prices of the inputs necessary for the extraction of coal remain unchanged. Inputs are assumed to be applied in fixed proportions in the exploitation of a deposit during a short-run period."[17] William Tabb studied microeconomic production functions for surface and underground mining in order to explain technological change and used data made available for individual mines. He found that a fixed coefficient technology due to the specialized nature of equipment and the resulting lack of substitutability was the best from his model.[18]

The operational time of any model is generally the time in which economic units can react to changes in market conditions. Consider a relationship of fixed proportions between the extraction of coal and the inputs. Algebraically,

$$Y_{ij} = \text{minimum} \, (X_{1ij}/b_{1ij}, \cdots X_{nij}/b_{nij}) \tag{3.1}$$

where Y_{ij} is the extraction of a unit of coal in the ith supply district by the jth method, X_{sij} is the level of the sth input, and b_{sij} is the input coefficient. The short-run extraction cost is described by the value of the factors used to extract a unit of coal,

$$C_{ij} = \Sigma b_{sij} p_{sij} \tag{3.2}$$

where the unit cost of Y_{ij} is C_{ij} and p_{sij} is the fixed unit price of the sth input. A capacity constraint to the output of coal from any supply district is assumed in this structure for the operational time period of our model. Assumption (3.1) implies that maximum production is not restricted by some inherent characteristic of the mine itself, but rather the level of an input. The limit to the available supply of this input can be a time lag before additional supply can be made available. We assume that the chief input involved in limiting supply for mining in a district is skilled labor which is available in a regular working year. Capacity for the extraction of Y_{ij} is therefore

$$K_{ij} = b_{1ij} \bar{X}_{1ij} W \tag{3.3}$$

where K_{ij} is the capacity of the district, b_{1ij} is the output of coal per man per day, $\bar{X}_{1ij}$ is the available skilled labor, and W is the number of days in a work year, which can include overtime. This implies that the model's time period is not sufficient to train new workers in response to regional output changes. It has often been noted that underground mining is unable to expand its labor force quickly.[19] Skills needed by underground miners are significant and take long training to acquire. Because of these skill requirements and the danger of accidents, many states require an experienced miner to accompany a new worker underground from six months to one year before allowing him to work alone.[20]

In surface mining, the limited input may well be capital equipment, and the labor skill involved is the ability to handle this large earth-moving equipment. Recently, a two-year backlog in the production of large draglines was reported.[21] According to Paul Bailly of Occidental Minerals, the time lag for large bulldozers and power shovels has advanced to three years, and no immediate solution appears near.[22] The highly specialized skills and equipment in surface mining allow little flexibility in mining operations or manning schedules. In order to mine 240 days per year, for example, certain key equipment must be operated for the entire period.[23] The operational time period of our model does not allow for the delivery of new surface-mining equipment.[24]

The basic model has 210 row constraints, the objective function, and 2,010 activities including a slack activity for each row constraint. The algebraic structure of the basic model is as follows:

$$\text{Minimize} \quad \sum_i \sum_j \sum_k \sum_l (C_{ij} + t_{il}) X_{ijkl} \tag{3.4}$$

such that

$$\sum_i X_{ijkl} \leqslant K_{ijk} \quad \text{for all } i,j,k \tag{3.5}$$

$$\sum_i \sum_j \sum_k B_{ijk} X_{ijk} \geqslant D_L \quad \text{for all } L \tag{3.6}$$

$$\sum_i \sum_j \sum_k S_{ijkl} X_{ijkl} \leqslant \bar{S}_L \quad \text{for all } L \tag{3.7}$$

$$X_{ijkl} \geqslant 0 \tag{3.8}$$

C_{ij} is the per-unit extraction cost of the jth mining method in the ith coal supply district, and t_{il} is the per-unit transportation cost from the ith district to the lth demand region.

X_{ijkl} is the coal extracted by the jth mining method in the ith district with the kth sulfur level and delivered to the lth demand region, and measured in 1,000 ton units.

K_{ijk} is the physical capacity of coal of the respective types in each district. In Appalachia, we also consider a slope-angle classification.

B_{ijk} is the energy value measured in Btus for the respective activities in millions of Btus.

S_{ijkl} is the average emitted sulfur per million Btus in the lth demand region, equivalent to $S_k b_{ijk}/D_l$, where D_l is the demand by utilities for steam electric coal measured in millions of Btus in region l, and S_k is the sulfur percent by weight of the kth sulfur class.

$\bar{S}_L$ ·represents the maximum allowable sulfur emission level. The five sulfur classes in the model, in sulfur emission per million Btu, are: 0-0.6, 0.6-1.0, 1.0-2.0, 2.0-3.0, 3.0 or greater.

The main behavioral assumption of the model is contained in the objective function (3.4). The optimal solution to (3.4) minimizes the total cost of the production, delivery, and utilization of steam electric coal subject to the constraint system. If a supply district can provide the requirements of a demand area at a lower delivered price than a competing district, either because of an extraction-cost or a transport-cost advantage, the model attempts by the specification in (3.4) to ship from that district. Utilities in each state are seen as attempting to minimize the costs of their fuel input deliveries, and producing districts are seen as shipping to the destinations which give them the greatest relative cost advantage. The minimum cost, optimal solution is the same that would occur in a perfectly competitive economy.[25]

The model is a set of interregional trade solutions with interaction due to relative levels of capacity and the diverse characteristics of natural resource deposits. Changes in the demand for a particular district's production can affect the regional distribution and utilization of the remaining areas' activities. The main factors involved in determining the minimum-cost solutions are the differences in extraction costs, transport costs, the spatial distribution of the demand, the characteristics of coal deposits, and the capacity limits. As indicated in (3.5), a short-run capacity constraint exists for production activities in a district. In such a model the comparative advantage of a district can result from the spatial distribution of demand and its interaction with transport costs, as well as production conditions. The objective function will be modified in later chapters as we consider reclamation costs and sulfur emissions taxes.

As seen in (3.6), each state must receive sufficient coal shipments to provide the energy values necessary to meet electric generating requirements for coal-fired plants. An assumption is implicit in (3.6) that the demand for energy supplied by coal in steam electric generation is independent of policy changes over the short run. There are four issues to be considered: (1) the demand for electricity by consumers; (2) the share of conventional steam electric generation; (3) the efficiency of steam electric generation; and (4) the proportion of conventional steam generation which is coal-fired. For (3.2) and (3.3) we argue that the time period of the model does not allow utilities to undertake major overhauls of existing plants or the construction of new ones.

Assume that the price of coal-fired generation increases as a result of some proposed policy. Over the long run, utilities could turn to the construction of central-station power plants using alternatives to steam, particularly nuclear plants. Similarly, cost increases in coal-fired generation could be offset by increasing the efficiency of plants and equipment so that coal with a lower heat rate could be used. Apparently this has occurred in oil and natural gas generation, particularly as their prices have increased.[26]

It is appropriate to place increases in the cost of coal in perspective when considering the possible price effects on electric demand. An increase in coal costs of 29 percent passed on fully to the electric-utility sector would increase the utilities' costs by only about 5.8 percent.[27] Inelastic price demands for electricity in the short run would tend to minimize these effects further.[28] If the cost of coal increases over the short run, the possibility of substituting another fuel is limited by the chronic short-run shortages in the supply of gas and oil for utilities and the time and cost needed to convert utility boilers to their use.

Since the heat rate of coal is less than other fossil fuels, an increase in the price of coal per million Btu represents a smaller relative increase in the cost of generating electricity. In particular, assume that all fossil fuel prices are the same. The data on heat rates in Chapter 2 suggest that an overnight increase in coal costs of approximately 10 percent per million Btu would not increase the cost of generating one kilowatt hour above that of oil or gas, but expectations of the future availability and price of the fuel would be a factor. Natural gas "curtailment schedules," recent oil shortages, the instability of fuel prices, etc., are considerations which would favor coal.* Our model allows us to consider these price changes in its dual problem.[29] The dual to our model is a maximizing problem which determines the delivered cost of coal-fired energy in each region. These costs, referred to as shadow prices in the programming literature, allow us to estimate the change in the supply price of coal in each region as a result of policy changes. Thus, we can determine the magnitude of any increase in the cost of coal in our analysis.

The basic model consists of the objective function (3.4) and the constraints (3.5), (3.6), and (3.8). The requirement in (3.8) is the usual nonnegativity constraint on the levels of activity in a linear programming problem.

The effects of policy through a sulfur-emissions standard can be introduced in (3.7), where the average sulfur emission per million Btu in each region is constrained to a particular level. The way in which this policy can cause regional effects is considered in Chapter 5.

We will relate our results in the following chapters to other studies that have considered the regional nature of coal production and policy impacts, though there have been few economic analyses of these issues. Given the programming structure of our model, we would comment here on our purpose relative to the general energy or air-quality model which is often a spinoff of the classical transportation problem.[30] The largest model which has a linear programming format was developed for EPA as an air-quality model.[31] This model,

---

*We will return to the issue of the competitive position of coal relative to our results in Chapters 8 and 9. We will also see the problems inherent in modeling the oil and natural gas sectors, particularly in our methodology.

which is intended to determine if air-quality programs are feasible, analyzes the criteria for air quality to be met at a "minimum acceptable selling price" (MASP) by fossil fuels, nuclear energy, etc.

Our analysis is concerned with the regional implications for land use and the effects of public policy in coal mining. Therefore, we have attempted to identify the policy issues, such as slope-angle control in Appalachia, that are important for land-use impacts and to determine their effects on coal on a regional basis. Rather than use concepts such as MASP or average value, we have attempted to construct detailed estimates of extraction costs and other regional variables. National models let total coal output grow at a specified rate, say 4 percent. Yet our main contention is that the growth rate of coal is more significantly affected by policy than that of any of our other natural resources. If we shift regional output patterns, can we necessarily assume that the same national rate of growth would occur over the long run? Few observers of the industry would minimize the importance of the effects of regional shifts.

Furthermore, the land-use impacts of coal can be significantly affected by policies that are not related to coal extraction or regional issues. In particular, we will emphasize the interaction between air quality proposals and regional land use. When we consider a tax on sulfur emissions for the purpose of air quality, for example, our analysis shows that two regions would benefit most. Though current capacity is used intensively, and the profitability of future mine investment is high, controversy over mining's impact on land use in these two areas is severe.

NOTES

1. For a complete description by counties, see United States Bureau of Mines, *Bituminous Coal and Lignite Distribution 1973* (Washington, D.C.: Government Printing Office, 1973), p. 3.

2. Phillip Giffin, *Industrial Concentration and Firm Diversification in Bituminous Coal*, Appalachian Resources Project (Knoxville, Tennessee: The University of Tennessee, 1972), p. 41.

3. For a discussion of the relatively complex regulatory issues involved in natural gas economics, see the two papers by James Hensen and Thomas Stauffer, "An Economic Rationale for Rationing Gas Supplies in the U.S." (unpublished paper, Harvard University, Cambridge, Massachusetts, 1973), and "The Rational Allocation of Natural Gas Under Chronic Supply Constraints" in *Energy: Demand, Conservation and Institutional Problems,* ed. Michael S. Macrakis (Cambridge: Massachusetts Institute of Technology Press, 1974), pp. 275-90.

4. Congressional Research Service, *Factors Affecting the Use of Coal in Present and Future Energy Markets* (Washington, D.C.: Government Printing Office, 1973), p. 4.

5. Ibid.

6. Richard Gordon, *U.S. Coal and the Electric Power Industry* (Baltimore: Johns Hopkins University Press, 1975), Chapter 2, p. 29.

7. See, for example, National Petroleum Council, *U.S. Energy Outlook,* December 1972, Chapter 5. More specifically, note the $1.50 per ton differential for "equivalent"

mines as derived from United States Bureau of Mines, *Basic Estimated Capital Investment and Operating Costs for Underground Bituminous Coal Mines,* IC 8632 (Washington, D.C.: Government Printing Office, 1974), and (updated) *Cost Analyses of Model Mines for Strip Mining of Coal in the United States,* IC 8535 (Washington, D.C.: Government Printing Office, 1972).

8. Computed from data in National Coal Association, *Bituminous Coal Data, 1973* (Washington, D.C.: National Coal Association, 1974), p. 15.

9. As reported in Federal Power Commission, *Monthly Report of Cost and Quality of Fuels for Steam Electric Plants* (Washington, D.C.: Government Printing Office, February 1973), p. 5.

10. Ibid., p. 6.

11. Leon Moses, "The General Equilibrium Approach," in Robert Dean et al., eds., *Spatial Economic Theory* (New York: Free Press, 1970), p. 15.

12. James Henderson, *The Efficiency of the Coal Industry* (Cambridge: Harvard University Press, 1958).

13. For a discussion of the transportation problem and its variants, see Robert Dorfman et al., *Linear Programming and Economic Analysis* (New York: McGraw-Hill Book Company, Inc., 1958).

14. United States Environmental Protection Agency, Office of Research and Development, *Environmental Considerations in Future Energy Growth,* Volume I, *Fuel-Energy Systems* (Washington, D.C.: Government Printing Office, 1973), p. 389.

15. Based on Executive Office of the President, Office of Emergency Preparedness, *The Potential for Energy Conservation* (Washington, D.C.: Government Printing Office, 1973), p. 36.

16. Gordon, op. cit., reports that some utilities experimenting with these coals have few conversion problems. See Chapter 7, *U.S. Coal and the Electric Power Industry.*

17. James Henderson, op. cit.

18. See William Tabb, *A Recursive Programming Model of Resource Allocation* (unpublished doctoral dissertation, University of Wisconsin, Madison, Wisconsin, 1968).

19. See Council on Environmental Quality, *Coal Surface Mining and Reclamation* (Washington, D.C.: Government Printing Office, 1973), pp. 55-59, and Congressional Research Service, op. cit., p. 28.

20. Council on Environmental Quality, op. cit., p. 59.

21. Hubert Hagen, "Regarding Proposed Surface Mine Legislation" in *Regulation of Surface Mining Operations,* 93rd Congress, 1st Session, Part 2 (Washington, D.C.: Government Printing Office, 1973).

22. Communication from Paul Bailly, President, Occidental Minerals Corporation, May 23, 1974.

23. See the manning schedules in United States Bureau of Mines, IC8535, op. cit. This study, occurring after Tabb's thesis, tends to verify his model.

24. We note that it has also been reported that the lead time for delivery of specialized underground equipment, such as cutters and ventilation systems, may take up to several years. See Council on Environmental Quality, op. cit., p. 57.

25. Moses, op. cit., p. 26. A more complete discussion is found in Louis Lefeber, *Allocation in Space: Production, Transport, and Industrial Location* (Amsterdam: North-Holland, 1958).

26. As noted in Chapter 2, the heat rate for coal is still lower than that for oil and natural gas.

27. See Schmidt-Bleek et al., *Benefit-Cost Evaluation of Strip Mining in Appalachia,* Appalachian Resources Project (Knoxville, Tennessee: The University of Tennessee, 1973), p. 21.

28. The largest estimates we have seen are the short-run price elasticity figures for the residential sector—0.14—in a study by T. Mount et al., *Electricity Demand in the United States: An Econometric Analysis* (Oak Ridge, Tennessee: Oak Ridge National Laboratories). Other studies find even smaller figures; see the summary of other studies on p. 11. In particular, there is debate whether there is even any long-run price elasticity.

29. For a discussion of the dual problem, see Dorfman, op. cit., Chapter 7.

30. For a review of such models, see Council on Environmental Quality, *Quantitative Energy Studies and Models* (Washington, D.C.: Government Printing Office, 1973), for a comprehensive survey. They consider both government and private institutions.

31. Their energy-quality model was developed by staff at Battelle Memorial Institute.

4

THE BASIC MODEL

The future role of coal in steam electric generation will depend critically on the availability of feasible fuel alternatives and the timing of new technological development. For the present and the 1980s, large additions to U.S. coal-fired capacity are under construction or are being planned. In particular, the additions begun in 1972 and 1973 will be on-line in 1978.

The late 1970s will be critical for decisions in electric utilities and the coal industry. Electric utilities will have to make firm decisions by that time on the kind and extent of increases in generating capacity which will meet the demand for electricity in the 1980s and 1990s. The decision to construct nuclear power plants using present technology must be made with sufficient lead time. At present it appears that about 10 years are required to plan and bring a light-water reactor plant into operation.[1] The corresponding planning period for a conventional steam electric plant is five years.

For the period after 1978, coal use could be increased through additional coal-fired capacity, but planning would have to begin almost immediately. The level of nuclear generation planned for the mid-1980s would be an important influence on the future of coal, depending on how technical and operating experience over the near term affected planned nuclear additions. Electric utilities, attempting to meet power demands with a reliable, inexpensive generating method, must consider the future costs and reliability of alternative fossil-fuel resources. Traditionally, coal has had the greatest domestic availability but public policies regarding safety, pollution, and reclamation could affect its desirability as a utility fuel.

Impending federal legislation dealing with reclamation requirements for land disturbed by mining activities will have a major effect on the coal industry. When this legislation becomes effective in 1978, it will be a permanent factor for consideration by mine operators.[2] There should be some initial resolution of uncertainty over sulfur-emissions policy and control technology by the 1980s.

Opening new mines may take four or five years because of the lead time needed to acquire capital equipment as well as to "perform geological analyses, develop mining plans, provide railroad spurs, and develop access roads and processing facilities."[3] A large surface mine in Appalachia producing one million tons per year requires an initial capital investment of approximately $13 million.[4] Even the smallest contour-surface-mine operator requires several hundred thousand dollars worth of equipment.[5] Because of uncertainty over the effects of public policy and doubts that large investments will earn adequate returns, mine operators may be reluctant to make the capital investment necessary to open new mines. The present "risky" nature of coal investments coupled with the large capital requirements may cause difficulties in getting the necessary funds from financial markets.[6]

The present policy will significantly affect the future of the Eastern and Western coal fields. In order to fulfill the generating requirements of utilities in the 1980s, mine investment plans will have to be made by 1978. Therefore, we will construct a simulation analysis of the projected steam electric coal market in 1978.

THE SOLUTION FOR 1978

The solution of the basic model is a simulation of the steam electric coal market in 1978, assuming a cost-minimizing system based on the spatial linear programming model described in the previous chapter. We assume that there is no public policy which limits slope angle, requires reclamation in surface mining, or attempts to limit sulfur emissions through taxes, standards, or bans on high-sulfur fuel. Regional variations in the characteristics of coal shipments are based on actual 1972 production and shipment data on the quality, cost, and source of the fossil fuels delivered to steam electric plants, which represent 98 percent of all delivered fuel.

There are two elements to consider in such a solution, i.e., the degree to which utilities attempt to minimize the costs of their fuel inputs and the competitive nature of the coal industry. The large utility company often has a purchasing department whose sole purpose is to find the cheapest sources of fuel. Given the importance of fuel input, this is not surprising. Gordon extends to the coal industry the argument of Morris Adelman that an alert demand sector is critical in providing a competitive setting.[7] He contends that the knowledgeable purchasing departments of these utilities could exploit the economic pressures and tensions that might exist in an oligopolistic situation.

On the supply side, the literature of industrial organizations dealing with coal indicates a general belief in the competititve nature of the coal industry.

TABLE 4.1

Regional Shipments by Mining Method: Basic Model
(millions of tons)

Region*	Surface	Underground	Total
Northern Appalachia	95.8	69.9	165.7
Central Appalachia	63.0	59.1	122.1
Southern Appalachia	14.6	5.5	20.1
Appalachia, total	173.4	134.5	307.9
Midwest	130.4	23.6	154.0
West	38.6	3.3	41.9
Total, United States	342.4	161.4	503.8

*The regions are identified as:
        Northern Appalachia = Districts 1-6
        Central Appalachia   = Districts 7,8
        Southern Appalachia = District 13
        Midwest   = Districts 9-12, 14, 15
        West   = Districts 16-23
Source: Compiled by the author.

Joe Bain wrote that, "the industry was and has remained extremely atomistic."[8] Bain's finding, often used as a guideline, is that an industry in which the eight largest sellers share 70 percent of the output enjoys profits greater than less concentrated industries. Phillip Giffin concludes that neither the national market nor the southeastern submarket would exhibit undesirable performance. Scherer concludes that the industry exhibits atomistic performance, except perhaps in the Midwest where large companies, such as Peabody Coal Company, are the major producers.[9] These considerations and the absence of monopoly profits may be explained by the contracts used by coal producers and electric utilities which have prices based on cost formulas.

As seen in Table 4.1, the basic model indicates that the general Appalachian area would be expected to be the major source of supply in 1978 under a situation constrained by a lack of policy as assumed for this beginning stage of our analysis. Northern Appalachia is the single largest producing area, followed by the Midwest and Central Appalachia. The importance of surface mining, accounting for approximately 68 percent of total shipments, is clearly seen. The distribution of shipments to major demand areas from these main supply regions is shown in Table 4.2. The influence of transportation costs on the interaction between the fixed location of the coal resources and the utility markets is also evident. The delivered cost advantage of supply districts in close proximity to each demand area is reflected in the solution.

TABLE 4.2

Regional Distribution of Coal
(thousands of tons)

FPC Power Regions	Supply Regions					
	Northern Appalachia	Central Appalachia	Southern Appalachia	Midwest	West	Total
New England						
Total tons	3,598.2	0	0	0	0	3,598.2
Percent of total	100	0	0	0	0	100
Middle Atlantic						
Total tons	65,119.2	0	0	0	0	65,119.2
Percent of total	100	0	0	0	0	100
Northeast Central						
Total tons	84,209.6	2,756.3	0	86,523.4	0	174,489.3
Percent of total	49	2	0	49	0	100
Northwest Central						
Total tons	0	0	0	24,112.2	17,432.3	41,544.5
Percent of total	0	0	0	58	42	100
South Atlantic						
Total tons	11,690.0	94,681.0	0	4,460.0	0	110,831.5
Percent of total	11	85	0	4	0	100
Southeast Central						
Total tons	0	24,616.1	20,142.0	38,892.6	0	83,650.7
Percent of total	0	29	24	47	0	100
Southwest Central						
Total tons	0	0	0	6,000.0	0	6,000.0
Percent of total	0	0	0	100	0	100
Western						
Total tons	0	0	0	0	24,483.3	24,483.3
Percent of total	0	0	0	0	100	100

Source: Compiled by the author.

THE ACTUAL DISTRIBUTION
OF STEAM ELECTRIC COAL

The primary purpose of our analysis in the following chapters is to suggest the magnitude of shifts in regional production which could be expected from alternative public policies, as compared to the unconstrained forecasts for regional production in 1978. Accordingly, since we are interested only in the resultant first differences from the 1978 estimates presented in Tables 4.1 and 4.2, the strict accuracy of the forecasts in those tables is not at issue. However, we do feel it necessary to check the consistency of our model by comparing the actual shipments of a particular production year with the predictions of our model.

Table 4.3 compares the regional distribution of steam electric coal in 1974 and our results. Our solution is based purely on cost minimization with assumptions of perfect certainty and information for both utilities and coal producers. Overall, we feel that we do approximate the actual pattern of the steam electric market, but there are reasons why we might expect some variations to occur. Appalachian underground producers have been concerned over the labor unrest which they have experienced since 1969.[10] Partly due to internal union strife and general discontent, work stoppages and wildcat strikes have occurred in addition to a long strike in 1971. Thus, utilities in areas where other sources of supply are able to compete with Central Appalachia may well balance a security premium with cost minimization. This may explain our tendency to overestimate Central Appalachian deliveries, particularly in the Southeast Central region.

Chicago has been using significant amounts of low-sulfur coal from Montana for electric generation, which Illinois and the Northeast Central area generally have not used. This special coal use in a major urban area explains why Western coal is shipped to Northeast Central utilities. Small percentages of shipments, such as 1-3 percent, may reflect old long-term contracts, special arrangements, etc., which we simply do not include in our model. The main characteristics of the utility market, however, are present in our model's solutions.

At this point we would like to make a comment concerning the transportation matrix in our study. Some linear programming models, in an attempt to reproduce actual patterns of distribution, limit their delineation of transportation activities from a supply source to a consuming source by the criterion of whether it is actually observed in the "real world." The model's solution obviously has no peculiarities, even though this is a result not so much of the validity of the model as it is of the way it was set up. In our model, over 60 percent of the possible activities which the model can consider are those in which actual distribution in 1972 and/or 1973 did not occur. This results from allowing

TABLE 4.3

Comparison of 1974 Percentage Regional Distribution of Coal and 1974 Estimates of Regional Distribution

FPC Power Regions	Supply Regions				
	Northern Appalachia	Central Appalachia	Southern Appalachia	Midwest	West
New England					
Actual	99	1	0	0	0
Predicted	100	0	0	0	0
Middle Atlantic					
Actual	97	3	0	0	0
Predicted	100	0	0	0	0
Northeast Central					
Actual	39	7	0	47	6
Predicted	49	2	0	49	0
Northwest Central					
Actual	0	2	0	60	38
Predicted	0	0	0	58	42
South Atlantic					
Actual	25	61	0	14	0
Predicted	22	68	0	10	0
Southeast Central					
Actual	0	18	17	65	0
Predicted	0	29	24	47	0
Southwest Central					
Actual	0	0	0	100	0
Predicted	0	0	0	100	0
Western*					
Actual	0	0	0	100	100
Predicted	0	0	0	0	100

*Including the Mountain and Pacific electric-power supply regions.

Source: Based on data from Federal Power Commission, Form 423, calendar year 1974, available on tape from Federal Power Commission, Washington, D.C.

Western states to ship over the entire state complex of the Midwest and East, a larger interaction between Midwestern and Eastern producers, etc. Thus, our results, compared to actual distribution patterns, are the model's solutions not forced solutions.

Table 4.4 makes the interesting comparison between the actual shares of regional production levels and the normative results of our model. We show a

TABLE 4.4

1973 Actual* and 1978 Estimated Pattern
of Steam Coal Extraction
(percent)

Region	1973	1978 Estimates
Appalachia	54.0	61.1
Midwest	34.4	30.6
West	11.6	8.3
United States	100.0	100.0

*1973 Actual from United States Bureau of Mines, *Mineral Yearbook, 1973* (Washington, D.C.: Government Printing Office, 1974), p. 65.
Source: Compiled by the author.

greater percentage of coal extraction in Appalachia for 1978 than at present. The share of Western and Midwestern coal is less. This result is readily explainable. The largest increases in coal-burning capacity which will come on-line by 1978 will be centered (in that order) in the South Atlantic, Northeast Central, Middle Atlantic, and Southeast Central power-supply regions.[11] These areas are exactly those markets which traditionally use Appalachian coal exclusively or at least heavily. This is reflected both in the actual 1973 distribution and our result for 1978. As widely expected, the use of Western coal is increasing in absolute terms over this time, but it is of interest to note that Western production does not increase in relation to the concentration of coal-fired capacity in the Appalachian market area. This location of coal-fired generation puts the districts of Appalachia in a particularly advantageous position, and our results reflect this.

THE ROLE OF TRANSPORT COSTS

The transport component of total delivered cost is particularly important for the Western producers, since they are the farthest from existing major markets. From 1966 to 1969, the average transport cost of coal shipments was relatively constant, due partly to the influence of low-cost unit-train operations and handling facilities designed primarily for coal shipments. In 1969, rail-freight rates for coal increased 14 percent across the board. Since then, the transport rates for rail-barge transportation have increased steadily. Consequently, we analyzed how a decrease in transport rates would affect supply districts, particularly those in the West, as compared with the solution in the basic model. This simulation was desirable because much of the discussion over the potential for Western coal is seemingly based on the transport rates in 1970 rather than those of 1974. The objective function (3.4) was therefore modified to the form of:

$$\text{Minimize } \Sigma \, \Sigma \, \Sigma \, \Sigma \, \big\{ [C_{ij} + t_{il} \, (1970)] X_{ijkl} + \phi t_{il} \, (1970) \, X_{ijkl} \big\}$$
$$\quad\quad\quad i \quad j \quad k \quad l$$

where t_{il} (1970) represents the transport rates of 1970 and ϕ is a parameter that allows us to simulate the increases in rates since that time.[12] The main results are given in Table 4.5. In Chapter 8 we will consider the issue of unit trains as a capacity limitation on coal shipments, particularly for Western coal. The Western producers were the only ones affected. The Appalachian districts, whether considering total shipments or relative changes in surface and underground production, were unaffected (meaning less than 0.5 million tons as a total change). Furthermore, the results were insensitive to the rates for values of rate increase up to $\phi = 0.40$, which tends to suggest the stability of the respective market areas of the supply districts that we will consider. For larger values, states such

TABLE 4.5

Western and Midwestern Shipment Changes under the Transport Cost Assumptions

Region	1970 Rates – (1 + ϕ) 1970 Rates,* thousands of tons	District with Largest Change
West		
Surface	−15,960.30	Montana
Underground	+913.87	Utah
Total	−15,046.43	
Midwest		
Surface	+368.00	Illinois
Underground	+12,084.22	Illinois
Total	+12,452.22	
United States		
Surface	−15,592.30	
Underground	+12,998.09	
Total	−2,594.21	

*$\phi \geqslant 0.4$.

Source: Compiled by the author.

as Wisconsin, which form the boundary of Western-Midwestern competition under the lower rates, can be supplied at lower delivered costs by Midwestern producers. The increase in Midwestern shipments reflects this delivered cost advantage as transport rates increase in only a few states, e.g., Wisconsin. The simulation aspect of the result can be noted by the fact that, in 1970, Western coal production was not sufficient to supply the "border" states of Western-Midwestern competition. Part of the reduction in Western surface production is the result of intra-Western competition. The higher transport costs allow the underground coal, which has a high heating value, to compete with Western surface-mined coal, which has a lower FOB cost but is more distant. This is why we see the increase in underground production in the Midwest.

DISCUSSION OF POLICY SIMULATIONS

In the following chapters, we will modify the structure of the basic model in order to analyze the effects of alternative environmental policies on the solution presented in this chapter. We will not only attempt to measure the impact of a particular policy alone, but to investigate how results can be altered if we consider the interaction of air-quality and land-use regulations. The solution in the basic model for 1978 will often be used as a benchmark to measure the regional impacts of a specific environmental policy, though we will also compare results among alternative levels of the same policy. In Chapter 5, the addition of a moderate sulfur emissions standard to the the model will be analyzed. The effects of two types of reclamation policies are analyzed in Chapter 6. The first policy considered requires reclamation of surface-mined land, which raises production costs but does not place any restriction on where mining may take place. The second policy prohibits surface mining on slopes of 20° or greater. In Chapter 7, a sulfur-emissions tax at four different rates is placed on regional coal utilization in the model. We will also consider the effects of a ban on the use of high-sulfur coal. In Chapter 8, to emphasize our conclusions on the interaction between sulfur emissions and the relative intensity of Midwestern-Western coal use, summary results of additional sulfur-emissions standards are given.

NOTES

1. Massachusetts Institute of Technology, Energy Policy Group, "Energy Self-Sufficiency: An Economic Evaluation," *Technology Review*, May 1974, pp. 42-43.

2. Environmental Policy Center, *Environment*, August 1974, p. 46.

3. Council on Environmental Quality, *Coal Surface Mining and Reclamation* (Washington, D.C.: Government Printing Office, 1973), p. 57.

4. United States Bureau of Mines, *Cost Analyses of Model Mines for Strip Mining Coal and Reclamation* (Washington, D.C.: Government Printing Office, 1973), p. 57.

5. This would provide for the costs of a "dragline, a coal loader, a bulldozer, and two trucks." See Council on Environmental Quality, op. cit., p. 57.

6. Conversation with Will G. Stockton, Vice President, Peabody Coal Company.

7. Richard L. Gordon, *U.S. Coal and the Electric Power Industry* (Baltimore: Johns Hopkins University Press, 1975), p. 69.

8. As reported in Phillip Giffin, *Industrial Concentration and Firm Diversification in Bituminous Coal*, Appalachian Resources Project (Knoxville, Tennessee: The University of Tennessee, 1972), p. 5. This dissertation summarizes the general literature on the "competitive" nature of coal.

9. Giffin, op. cit., p. 5. A similar conclusion is reached by Reed Moyer, *Competition in the Midwestern Coal Industry* (Cambridge: Harvard University Press, 1964).

10. Energy Policy Project of the Ford Foundation, *Exploring Energy Choices* (Washington, D.C.: Ford Foundation, 1974), pp. 7, 36. Also see Richard Gordon, "Coal—Our Limited Vast Fuel Resource" in E. Erickson et al., *The Energy Question, Volume 2* (Toronto: University of Toronto Press, 1974), p. 73.

11. See Executive Office of the President, Office of Emergency Preparedness, *The Potential for Energy Conservation* (Washington, D.C.: Government Printing Office, 1973), B-2.

12. The values for ϕ in our relative simulation were 0, 0.2, 0.4, 0.6, 0.96. The approximate rate relative to 1970 as used in our study would be 0.77, based on the rate data supplied and suggested by Robert Young.

5

THE SULFUR STANDARD

The solutions in the previous chapter did not allow for the role of sulfur emissions as an important variable in coal use. We have discussed the use of the regulatory standard as a sulfur-emissions policy tool and seen that it can be introduced by adding a set of constraints for utility emissions to the model. The transformation of the sulfur oxide emissions is not a weight-for-weight reaction, that is one pound of sulfur in coal is turned into 1.9 pounds of sulfur oxides.[1] The average sulfur content in coal, as shipped to electric utilities in 1972, contained 2.9 percent sulfur, as shown in Table 5.1. Yet this average, including the low value of Western coals, is clearly exceeded in many districts, even some which are major producing areas such as Western Kentucky and Illinois. Moreover, since their coals have no coking or related properties, these districts produce almost entirely for electric utilities. For example, 94 percent of all production in Western Kentucky was delivered to utilities in 1973.[2] Similarly, the southwestern district, composed almost entirely of Missouri area strip mines, delivered 94 percent of its total production to utilities in 1972, even though its coal had the highest average sulfur content in the continental United States. The Bureau of Mines is beginning to report Texas lignite production in this district, but it also is primarily for utility use.

Our purpose was not to examine a complex system of regulations, but rather to investigate the regional impacts that would be likely to occur if we assume a policy that discourages the use of high-sulfur emissions. The lowest secondary standard considered by EPA in 1971 for new coal-fired plants was a level of 1.2 pounds of sulfur oxides per million Btu, or about 0.6 pounds of sulfur per million Btu. These were based on the optimistic expectations of completely operational stack-gas scrubbing systems with up to 90 percent efficiency rates. As noted in the previous chapter, if the first systems to appear

TABLE 5.1

Regional Shipments of Steam Electric Coal by Average Sulfur Content

District of Origin		Average Sulfur Content		Quantity Shipped, thousand tons	
		1972	1973	1972	1973
1	Eastern Pennsylvania	2.2	2.2	34,346	34,362
2	Western Pennsylvania	2.1	2.1	8,923	9,403
3 and 6	Northern West Virginia and panhandle	2.7	2.7	36,836	34,818
4	Ohio	3.5	3.5	41,731	38,926
7	Southern, numbered 1	–	–	746	789
8	Southern, numbered 2	3.9	3.9	71,313	65,747
9	Western Kentucky	4.6	4.6	49,374	52,894
10	Illinois	3.4	3.4	53,137	49,705
11	Indiana	3.4	3.4	20,286	20,454
12	Iowa	3.4	3.4	703	618
13	Southeastern	1.7	1.7	13,049	11,628
14	Arkansas-Oklahoma	–	–	–	–
15	Southwestern	4.9	4.9	7,430	12,665
16	Northern Colorado	0.5	0.5	561	492
17	Southern Colorado	0.6	0.6	2,566	2,974
18	New Mexico	0.6	0.6	10,184	11,008
19	Wyoming	0.6	0.6	10,637	14,113
20	Utah	0.7	0.7	1,611	1,903
21	North and South Dakota	0.9	0.9	6,032	6,098
22 and 23	Montana and Washington	1.5	1.5	10,797	13,567

Source: 1972 data are from United States Bureau of Mines, Coal–Bituminous and Lignite in 1972 (Washington, D.C.: Government Printing Office, 1973), pp. 65, 71. The 1973 data are from National Coal Association, Bituminous Coal Data, 1973 (Washington, D.C.: National Coal Association, 1974), pp. 83, 85.

are the double alkali or Wellman-Lord, an efficiency of 90 percent may be optimistic. Hopes for these standards in the near future are limited. The electric utilities have contested any state environmental agency action that has attempted to impose even modest emissions standards. These suits have worked their way through lower courts and are beginning to reach the appeals courts. The lower courts, hearing cases on the relation of sulfur emissions to ambient standards, ambient standards to specific sources, specific sources to available technology, etc., have generally supported the utilities' position for moderate standards.[3] One of the more important cases has recently been decided in Pennsylvania. The Pennsylvania Commonwealth Court has unanimously upheld a lower court opinion that "emission control technology is not commercially available for the removal of sulfur oxides from stack gases."[4] Studies of sulfur emissions, particularly in New York, have often calculated emissions in a regulatory atmosphere of 1 percent average sulfur content, even assuming the implementation of the Clean Air Act.[5] This is usually obtained from a mixture of lower sulfur oil and higher sulfur coals. The corresponding sulfur emissions per million Btu would depend on the Btu value per ton of coal, but emitted sulfur for lower-rank coals could exceed 1.25 pounds per million Btu.

Considering the sulfur data in Table 5.1 and the respective district production, we utilize constraints limiting the average sulfur content of coal to 2 pounds of sulfur per million Btu in order to simulate the effects of a standard that discourages the use of high-sulfur coal.* The resulting regional shipments and deliveries, given in Table 5.2, conflict with the generally accepted hypothesis that Eastern coal producers would be severely hurt by sulfur restrictions and that Western producers would benefit from controls. When we compare the differences between the basic model and the solution for a policy that discourages high sulfur use, we find that Central Appalachia, not the West, increases output most.

Table 5.3 compares the regional deliveries under the two scenarios. An explanation for this seemingly contradictory result lies in the nature of Western and Central Appalachia coal production. Approximately 62 percent of all production in Central Appalachia has a sulfur content of less than 1 percent. Indeed, 97.3 percent of this area's coal production has a sulfur content less than or equal to 2 percent.[6] In the basic model, much available low-sulfur underground coal in this area is not used since it has a high production cost compared to surface-mined coal near major markets in the East. Yet a sulfur regulation would make this coal quite desirable, and we find that underground production would increase by 16 million tons under such a regulation. Indeed, surface-mined coal actually decreases in Appalachia and in the United States as a whole, since

---

*The attempt to lower the average standard to 1.3 pounds of emitted sulfur per million Btu results in unacceptable solutions. We will return to this point in Chapter 9.

TABLE 5.2

Regional Shipments with a Sulfur Emissions Standard of 2.00 pounds
(millions of tons)

| | Mining Method | | |
	Surface	Underground	Total
Northern Appalachia	92.7	74.7	167.4
Central Appalachia	63.0	75.1	138.1
Southern Appalachia	14.6	5.5	20.1
Appalachia, total	170.3	155.3	325.6
Midwest	120.1	29.1	149.2
West	45.6	3.4	49.0
Total, United States	336.0	187.8	523.8

Source: Compiled by author.

TABLE 5.3

Comparative Changes in Regional Shipments from a Sulfur Emissions Standard of 2.00 pounds to no Standard
(millions of tons)

| | Mining Method | | |
Region	Surface	Underground	Net Change
Northern Appalachia	−3.06	+4.77	+1.71
Central Appalachia	0	+16.05	+16.05
Southern Appalachia	0	0	0
Appalachia, total	−3.06	+20.82	+17.76
Midwest	−10.30	+5.51	−4.79
West	+7.01	0	+7.01
Total, United States	−6.35	+26.34	+19.99

Source: Compiled by author.

64

underground coal would replace some high-sulfur surface-mined coal. The surface-mined deliveries from the West increase by 18.3 percent, up by 7 million tons.

It is the low-sulfur production of Central Appalachia which allows the sulfur constraint to be met. The increase in Western production is sent to the states of the Northeast Central and Northwest Central regions, roughly identifiable as the Midwest. By mixing this additional low-sulfur coal with high-sulfur coal from the nearby Midwest utilities can still meet the standard. Western coal does not penetrate Eastern markets and our results indicate that future markets for Western coal will continue to be mainly in the Midwest and not the East.[7] The relatively low heating value of Western coal is shown by the fact that a decrease in Midwestern coal production of approximately 5 million tons requires an increase in Western shipments of 7 million tons, a 40-percent increase in total tonnage. Indeed, the required surface mining formerly in the Midwest has been transferred to the Western fields with an additional increase in total tonnage. The increase in total underground production is around 17 percent, as lower-sulfur coal use is stimulated. Central Appalachia has the largest increase, 27 percent.

The regional impact of stimulating underground production in Central Appalachia can be important. Per-capita income in this area is the lowest in Appalachia and only about half the U.S. average.[8] It has the smallest manufacturing base and is the only Appalachian area showing a decline in population during the last decade (of 10.7 percent). The real unemployment rate in 1970 was 18 percent. The average daily wage, including all fringe benefits, has been estimated in 1973 for a unionized underground miner to be $65.50.[9] Any policy which would stimulate the outlook for underground mining would have a beneficial effect on this regional economy.

The total emission of sulfur is significantly reduced by the standard. In the basic model, 11.41 million tons of sulfur are emitted, representing almost

TABLE 5.4

Changes in Steam Coal Costs
(billions of dollars)

Total	Without Standard	With Standard	Net Change
Extraction	3.634	3.657	+0.023
Transportation	1.330	1.451	+0.121
Total	4.964	5.108	+0.144

Source: Compiled by author.

TABLE 5.5

Comparison of the Changes in the Delivered Costs of Steam Coal with the Sulfur Emissions Standard

State	Dollars* per Million Btu		
	Without Standard	With Standard	Net Change
Alabama	0.49	0.49	0.00
Arizona	0.36	0.36	0.00
Arkansas	—	—	—
California	—	—	—
Colorado	0.40	0.40	0.00
Connecticut	0.60	0.64	0.04
Delaware	0.50	0.52	0.02
District of Columbia	0.52	0.55	0.03
Florida	0.60	0.64	0.04
Georgia	0.49	0.53	0.04
Idaho	—	—	—
Illinois	0.45	0.67	0.22
Indiana	0.40	0.61	0.21
Iowa	0.50	0.55	0.05
Kansas	0.50	0.55	0.05
Kentucky	0.39	0.50	0.11
Louisiana	—	—	—
Maine	—	—	—
Maryland	0.52	0.56	0.04
Massachusetts	0.63	0.67	0.04
Michigan	0.54	0.59	0.05
Minnesota	0.59	0.59	0.00
Mississippi	0.54	0.61	0.07
Missouri	0.42	0.66	0.24
Montana	0.31	0.31	0.00
Nebraska	0.49	0.49	0.00
Nevada	0.52	0.52	0.00
New Hampshire	0.64	0.68	0.04
New Jersey	0.57	0.60	0.03
New Mexico	0.30	0.30	0.00
New York	0.58	0.61	0.03
North Carolina	0.49	0.52	0.03
North Dakota	0.40	0.40	0.00
Ohio	0.48	0.59	0.11
Oklahoma	0.49	0.54	0.04
Oregon	—	—	—
Pennsylvania	0.44	0.46	0.02
Rhode Island	0.63	0.65	0.02
South Carolina	0.53	0.56	0.03
South Dakota	0.53	0.53	0.00
Tennessee	0.39	0.42	0.03
Texas	0.54	0.54	0.00
Utah	0.39	0.39	0.00
Vermont	0.63	0.67	0.04
Virginia	0.52	0.55	0.03
Washington	—	—	—
West Virginia	0.41	0.44	0.03
Wisconsin	0.52	0.69	0.17
Wyoming	0.31	0.31	0.00

*1972 dollars.
Source: Compiled by author.

22.82 million tons of sulfur oxide pollutants. With the emissions standard, total sulfur emissions are reduced by almost 8 percent, to a level of 10.64 million tons. This decreases the corresponding amount of sulfur oxide pollutants by 1.83 million tons.

The changes in total steam coal costs from the sulfur emissions standard are indicated in Table 5.4. Total costs increase by 3 percent, or $144 million. Of this, 2.5 percent is due to the 9-percent increase in transportation costs caused by the increased use of Western coal. This is expected, since Western deposits are beginning to replace local production in the Midwest market.

The direct cost of the sulfur-emissions abatement achieved by the standard is the change in the regional supply prices paid by utilities for coal. The shadow prices associated with the constraints on utility demand give us the magnitude of the increase in coal prices. These price changes are not regionally uniform, since the manner in which the sulfur emissions regulation is met differs, as seen in Table 5.5. The states in the Midwest experience the largest price increases because of higher transport costs of Western coal used to replace Midwestern shipments. The additional distance component dominates any decline in extraction costs in the Western fields. States in the South and Middle Atlantic areas, generally identified as the East or Southeast, have seen little change, since much of the coal used in a no-sulfur-policy model dependent on relative costs and heating value alone comes from Central Appalachia in the first place. Similarly, the states in the West are relatively unaffected. There is an increase in Kentucky because high-sulfur coal surface mined in western Kentucky is replaced by lower-sulfur underground coal from the eastern part of the state. The conclusion here is that the Midwest would experience the biggest change in delivered coal prices, approximately 22 cents per million Btu on the average. Traditionally, only the effects on the East, not the Midwest, have been considered.

The dual programming problem to our regional minimization problem determines the quasi rent of the supply capacity of each district. These quasi rents, i.e., the amount a particular deposit earns above its costs, have been key variables in planned economies which model their systems in a programming format.[10] Each quasi rent can be interpreted as how much the total costs of the system's solution would be reduced if the respective capacity was increased enough to allow the production of an additional unit of coal. By studying the manner in which these quasi rents vary as a function of simulated changes in policy, planners can determine the location of investment in new capacity. For our purposes they also suggest the regions in which we might expect the greatest incentive to add mine investment under different policies. Capacities which are not used completely earn no quasi rent, and the delivered price in any state is set by the marginal supply source. The delivered price of the least efficient source of steam electric coal in each respective state will set the costs of coal use in that state. The capacities used from other sources will earn a quasi rent as a result.

Table 5.6 indicates how the sulfur emissions standards affect the relative regional quasi rents. Sulfur emissions standards place a premium on low sulfur production. Underground production of lower-sulfur coal, which was previously

TABLE 5.6

Comparative Quasi Rents with and without the Sulfur Emissions Standard
(dollars per ton of capacity)

	Sulfur Level, pounds per million Btu				
Region	0-0.6	0.6-1.00	1.00-2.00	2.00-3.00	3.00 or greater
Central Appalachia					
Without Standard					
Surface	1.07	0.69	0.73	0.48	0.63
Underground	0.19	0.00	0.15	0.03	0.00
With Standard					
Surface	2.45	1.36	1.40	1.14	1.29
Underground	1.62	0.69	0.85	0.73	0.43
Midwest					
Without Standard					
Surface	*	0.88	0.42	0.73	0.30
Underground	0.01	0.63	0.63	0.11	0.00
With Standard					
Surface	*	3.72	2.10	1.23	0.00
Underground	3.59	2.51	2.51	0.62	0.00
West					
Without Standard					
Surface	1.98	2.18	2.54	*	*
With Standard					
Surface	1.98	2.18	2.54	*	*

*No such coal produced.
Source: Compiled by author.

at a cost disadvantage, earns the largest increase. The large changes in the Midwest reflect the delivered cost advantage that local low-sulfur production gains over Western coal. In Central Appalachia, it is of interest to recall that the lower-sulfur deposits earning the highest quasi rents generally occur on production slopes exceeding $20°$. Since cost considerations alone compel Western utilities to use coal from the same region, whether or not a standard is in force, the relative return stays the same.

Our main conclusion regarding the simulated effects of enforcing a sulfur emissions regulation is that the regional impacts are somewhat contrary to what was expected. The Midwest, not the Appalachian area, suffers the largest impact, experiencing the greatest fall in production as well as the largest price increases in electric generation. Indeed, contrary to the notion that Western surface production would increase most, Central Appalachia would be in the best position of all areas considered if sulfur emissions are regulated. This points out the interaction between markets and supply districts when a spatial dimension is recognized. The fields of low-sulfur Central Appalachian coal are far closer to utilities in the East than the surface production in the West. The Western fields increase their production more due to the relatively high-sulfur values of Midwest producers than from any delivered cost advantage.

NOTES

1. United States Environmental Protection Agency, *Compilation of Air Pollution Emission Factors* (Research Triangle Park, North Carolina: Environmental Protection Agency, 1973), pp. 1-3.

2. Derived from National Coal Association, *Bituminous Coal Data, 1973* (Washington, D.C.: National Coal Association, 1974), p. 85.

3. The reader interested in a running commentary of such cases should see the industry magazine, *Coal Age,* particularly the section, "This Month in Coal."

4. *Coal Age*, May 1974, p. 27.

5. See Q. Hausgaard, "Proposed Tax on Sulfur Content of Fossil Fuels," *Public Utilities Fortnightly* 88, no. 6 (1971): 27-33; Duane Chapman et al., "Electricity and the Environment," paper for AAAS meeting, Philadelphia, December 26, 1971; Chapman, "Internalizing an Externality: A Sulfur Emission Tax and the Electric Utility Industry," *Energy: Demand, Conservation and Institutional Problems* (Cambridge: Massachusetts Institute of Technology Press, 1974), pp. 190-208.

6. Based on Federal Power Commission, Form 423 data, 1973; these data form a main source in our study.

7. See "New Look at Western Coal," *Coal Age,* May 1974, pp. 75-130.

8. The figures in this discussion are from Council on Environmental Quality, *Coal Surface Mining and Reclamation* (Washington, D.C.: Government Printing Office, 1973).

9. Massachusetts Institute of Technology, Energy Policy Group, "Energy Self Sufficiency: An Economic Evaluation," *Technology Review,* May 1974, p. 38.

10. This use is discussed in Leon Moses, "The General Equilibrium Approach," in Robert Dean et al., eds., *Spatial Economic Theory* (New York: Free Press, 1970), p. 26.

6

ISSUES IN RECLAMATION

In this chapter, we analyze the effects of proposed reclamation policies after a discussion of regional issues in reclamation and its costs. We consider two types of policy. The first requires reclamation of surface-mined land after the mining operation has ceased. We will concentrate on the requirements that the original topography be restored "back to contour." The second policy prohibits contour surface mining on slopes greater than 20° and is explicitly designed to control the environmental problems of surface mining in Appalachia.

The bituminous coal industry utilizes more land in the United States than any other mining industry.[1] Bituminous coal was responsible for 40 percent of land utilized by mining over the period 1930-71, with its relative share in 1971 at 35 percent. Therefore, since other mining industries utilize more land than the coal industry, it is evident that land reclamation is not an issue applicable to coal extraction alone. Indeed, as seen in Table 6.1, the bituminous coal industry reclaimed more land than any other over the 1930-71 period, with 68 percent reclamation of land utilized. By today's standards, however, this reclamation may be viewed as inadequate or too varied across or even within states.

As seen in Table 6.2, the impact of mining activities on land use differs across states. Sand and gravel operations form the second largest sector with 22 percent of land used in 1971, or approximately 63 percent as much land as is used in coal mining. The increased pressure for reclamation in mining activities can be seen by comparing the 75 percent reclamation of sand and gravel operators in 1971 with the average of around 30 percent in the period 1930-71. The coal industry, supplemented by the work of other state and environmental organizations, actually reclaimed more acreage in 1971 than was utilized in mining activities.

TABLE 6.1

Land Utilized and Reclaimed by the Mining Industry in the United States in 1930-71 and 1971, by Selected Commodity

Commodity	Land utilized, acres[a]		Land Reclaimed, acres[a]		Percent Reclaimed, 1930-71
	1930-71	1971	1930-71	1971	
Metals					
Copper	166,000	19,100	4,810	1,410	2.9
Iron ore	108,000	8,620	4,630	2,330	4.3
Uranium	12,800	1,950	810	440	6.3
Other[b]	237,000	6,740	33,000	8,400	13.9
Total[c]	524,000	36,400	43,300	12,600	8.3
Nonmetals					
Clays	167,000	7,460	58,700	4,330	35.1
Phosphate rock	77,300	10,200	12,300	2,070	15.9
Sand and gravel	660,000	46,400	197,000	34,300	29.8
Stone	516,000	25,000	124,000	9,480	24.0
Other[d]	138,000	6,030	14,100	3,070	10.2
Total[c]	1,560,000	95,100	406,000	53,200	26.0
Solid fuels					
Bituminous coal	1,470,000	73,200	1,000,000	94,600	68.0
Other[e]	105,000	1,710	14,100	2,230	13.4
Total[c]	1,570,000	74,900	1,010,000	96,900	64.3
Grand total[c]	3,650,000	206,000	1,460,000	163,000	40.0

[a] Includes area of surface-mine excavation, area used for disposal of surface-mine waste, surface area subsided or disturbed as a result of underground workings, surface area used for disposal of underground waste, and surface area used for disposal of mill or processing waste.

[b] Bauxite, beryllium, gold, lead, manganese, mercury, molybdenum, nickel, platinum-group metals, silver, titanium (ilmenite), tungsten, vanadium, and zinc.

[c] Data may not add to totals shown because of independent rounding.

[d] Aplite, asbestos, barite, boron minerals, diatomite, emery, feldspar, fluorspar, garnet, graphite, greesand marl, gypsum, kyanite, lithium minerals, magnesite, mica, millstones, olivine, perlite, potassium salts, pumice, pyrites, salt, sodium carbonate, talc, tripoli, vermiculite, and zeolite.

[e] Anthracite and peat.

Note: Mining of bituminous coal accounts for 40 percent of land utilized in 1930-71, and 35 percent in 1971; sand a gravel operations for 18 percent, and 22 percent in 1971; and crushed and broken stone for 14 percent, and 12 percent in 1971. Land use for copper and clay production was nearly equal for the 42-year period and was 166,000 acres and 167,000 acres, respectively; however, land utilization for copper in 1971 was about 2.5 times greater than that for clay. Collectively, nonmetals utilized the same quantity of land as bitimous coal and antracite in the 42-year period; however, in 1971, nonmetals exceeded coal in quantity of land used.

Source: United States Bureau of Mines, *Land Utilization and reclamation in the Mining Industry, 1930-1971*, IC8642 (Washington, D.C.: Government Printing Office, 1974), p. 54.

TABLE 6.2

Land[a] Utilized and Reclaimed by the Mining Industry in the United States in 1971, by State and Commodity Group
(acres)

State	Metals		Nonmetals		Fossil Fuels[b]		Total[c]	
	Utilized	Reclaimed	Utilized	Reclaimed	Utilized	Reclaimed	Utilized	Reclaimed
Alabama	180	170	1,000	590	3,170	2,160	4,360	2,930
Alaska	130	380	680	490	210	750	1,020	1,630
Arizona	10,900	670	2,260	650	W	W	13,200	1,320
Arkansas	140	80	1,120	900	150	170	1,420	1,150
California	1,760	2,060	10,300	6,610	W	W	12,000	8,670
Colorado	300	340	4,460	970	380	470	5,140	1,780
Connecticut	–	–	540	400	–	–	540	400
Delaware	–	–	200	60	–	–	200	60
Florida	890	120	11,000	2,240	20	4	11,900	2,360
Georgia	410	180	1,930	1,030	1	1	2,340	1,210
Hawaii	–	–	330	80	–	–	330	80
Idaho	360	1,000	1,070	560	W	W	1,430	1,560
Illinois	W[d]	W	3,710	2,410	7,600	12,900	11,300	15,300
Indiana	–	–	2,090	1,340	4,350	8,030	6,440	9,380
Iowa	–	–	1,620	1,320	120	420	1,740	1,740
Kansas	–	240	1,170	760	W	W	1,170	1,010
Kentucky	W	W	950	540	13,500	14,100	14,500	14,500
Louisiana	W	W	1,230	670	–	–	1,240	670
Maine	W	W	650	580	50	50	700	630
Maryland	–	–	1,210	740	250	370	1,460	1,110
Massachusetts	–	–	1,450	850	W	W	1,450	850
Michigan	960	180	3,840	2,660	100	150	4,900	3,000
Minnesota	5,400	1,640	2,130	1,550	W	W	7,540	3,190
Mississippi	–	–	700	410	–	–	700	410

	36,400	12,600	95,100	53,200	74,900	96,900	206,000	163,000
Missouri	W	W	1,780	1,030	W	W	1,780	1,030
Montana	1,510	270	2,070	1,430	740	460	4,320	2,170
Nebraska	—	—	940	620	—	—	940	620
Nevada	2,190	790	950	440	—	—	3,140	1,230
New Hampshire	—	—	330	240	—	—	330	240
New Jersey	W	W	1,180	970	60	4	1,240	970
New Mexico	2,790	330	1,000	430	W	W	3,790	750
New York	360	150	2,330	2,030	20	—	2,710	2,180
North Carolina	W	—	2,240	960	—	—	2,240	960
North Dakota	—	1	550	420	1,000	1,590	1,500	2,010
Ohio	—	—	3,450	2,740	9,560	13,000	13,000	15,800
Oklahoma	W	W	690	490	W	W	690	490
Oregon	20	320	1,620	980	—	1	1,640	1,300
Pennsylvania	110	40	2,510	1,260	11,800	17,900	14,400	19,200
Rhode Island	—	—	180	80	—	—	180	80
South Carolina	—	—	790	510	10	—	800	510
South Dakota	80	20	910	700	—	10	1,000	740
Tennessee	W	W	1,560	1,120	2,070	2,050	3,630	3,170
Texas	360	250	4,830	2,500	W	W	5,190	2,750
Utah	3,140	850	990	640	W	W	4,130	1,490
Vermont	—	110	250	120	—	—	250	240
Virginia	W	W	2,260	1,100	2,900	2,830	5,160	3,930
Washington	60	60	2,000	1,380	W	W	2,060	1,440
West Virginia	—	—	500	360	11,100	12,500	11,600	12,900
Wisconsin	W	W	2,130	1,540	W	W	2,130	1,540
Wyoming	1,690	190	1,510	740	750	540	3,940	1,460
Undistributed	2,640	2,140	—	—	5,230	6,440	7,870	8,590
Total[c]	36,400	12,600	95,100	53,200	74,900	96,900	206,000	163,000

[a] Includes area of surface mine excavation, area used for disposal of surface mine waste, surface area subsided or disturbed as a result of underground workings, surface area used for disposal of underground waste, and surface area used for mill or processing waste.

[b] Excludes oil and gas operations.

[c] Data may not add to totals shown because of independent rounding.

[d] W: Withheld to avoid disclosing individual company confidential data, included with "Undistributed."

Source: United States Bureau of Mines, *Land Utilization and Reclamation in the Mining Industry, 1930-1971*, IC8642 (Washington, D.C.: Government Printing Office, 1974), p. 13.

The current controversy over the land use by the mining industry, the focus of this chapter, is centered specifically on the extraction of coal and its effects. Reclamation of the large open pits in metal industries, particularly those in copper and iron ore mining, is not widespread, only 8.3 percent as seen in Table 6.1. Some pits are still being mined and other inactive pits are maintained as potential sources for lower-grade ores. Many of the exhausted pits serve as water reservoirs for local communities or industry. These large pit excavations are difficult to reclaim because of their great depth.

Adequate reclamation and soil stability, particularly on steep slopes, have been the primary causes of concern in the producing areas of Appalachia. Kentucky has generally been considered as the model state for surface-mining legislation, but even after 1964, when reclamation laws were passed to control the amount of spoil thrown over the slope, landslides continued to occur. Field data collected in a three-year research program conducted by Kentucky's division of reclamation showed that 86 percent of the landslides occurred on slopes exceeding 20°.[2] In an experimental effort to reduce landslides in 1970, Kentucky asked mine operators to spread the overburden over the downslope to prevent landslides. Because of widespread erosion and relatively high rainfall in this area, however, landslides continued, and it was decided that these experiments could "reduce neither the amount of environmental damage nor the number of operator violations."[3]

There is also concern over the long-run impact of scarred, unreclaimed surface-mining lands. William Miernyk argued that improving the environmental quality of the Appalachian area would increase the potential for "amenity-oriented economic growth," that is, regional growth from increased tourism and the in-migration of skilled laborers who place a high value on their physical environment.[4] The Appalachian Regional Commission has allocated funds for highways linking the Appalachian area and the rest of the Eastern states. The Commission feels that the wilderness of Appalachia will attract these amenity-oriented economic activities.[5] This new growth potential would be hindered if surface-mined lands were not reclaimed.

In the Midwest the environmental issues in strip mining deal mainly with the elimination of unreclaimed land from alternative uses in the future and the disruption of wildlife habitats. The coal industry often points out the potential of reclaimed lands for grazing, farming, orchards, and possibly new wildlife and lake areas.[6]

Revegetation, establishing new ground cover on disturbed mining lands, is an accepted method of reclamation in both the wet, humid regions of Appalachian coal production and the mining areas of the Midwest. The issue in these regions is the type and extent of revegetation that should be required. Should mine operators, attempting reclamation, be allowed to use only fast growing, single-species trees or should they be required to plant diverse species of trees and shrubs, including berries, etc., which would attract new wildlife? Another question in revegetation is the ability, on a short-term basis, of nurseries to supply the needed seedlings quickly. In the Eastern coal-producing areas, almost 200 million seedlings would be required annually for revegetating currently mined areas. In Tennessee, for example, out of 36 potentially adequate species of trees and shrubs only three are actually available.[7]

In the relatively dry areas of the West, however, there is some question of whether proper revegetation is possible as a method of reclamation. Soil conditions in the West are extremely fragile and direct rainfall and other sources of water are limited.[8] A study undertaken by the National Academy of Sciences concludes that reclamation does not appear feasible where the rainfall is less than 10 inches annually and where soils have difficulty in retaining the moisture.[9] These minimum conditions are met in the two most heavily mined areas in the West, the Northern Great Plains and the Rockies, which contain 60 percent of the region's surface-mine reserves of coal. However, the study concludes that, although water requirements for mining and reclamation in these areas can be met, "there is not enough water available there for large scale operations like gasifying and liquefying coal or generating electric power."[10] The implication of the report for reclamation in other Western areas is not clear, since it emphasizes revegetation issues.

The remaining Western area of significant surface production is found in the generally contiguous fields of northeastern Arizona and northwestern New Mexico.[11] The production of this area is used mostly by Western utilities, particularly those in Arizona. In parts of these states, annual rainfall can be less than 5 inches and the surface soil is generally alkaline. As a result, little or no vegetation exists under normal circumstances. Since most of the deposits currently under extraction are on Navajo or Hopi Indian lands, the effort at present is to determine which vegetation will produce the best grazing crop for sheep and cattle.[12]

REGIONAL COSTS OF RECLAMATION

Estimating surface-mining reclamation costs is difficult because costs can vary according to (1) the level of reclamation to be required, (2) the site-specific geological conditions of the mine, and (3) whether or not the reclamation is carried out as an integral part of the mining operation. Furthermore, if revegetation is needed, cost can be affected by the amount of moisture in the area, soil conditions, and the type of material in the surface layer of the soil.

Reclamation costs differ between contour strip mining and area mining, for it is more difficult to restore a hillside or mountain slope than to grade and revegetate the spoil in relatively flat areas. The cost of reclamation in area strip mining, particularly on a per-ton basis, varies between the Midwestern and Western fields, primarily according to the thickness of coal seams and the amount of overburden required to regrade. Coal seams can run 25 to 100 feet thick in the main Western fields of Montana and Wyoming, with only 50 to a few feet of overburden, while seams in the Midwest are generally only 3 to 6 feet thick with an average overburden close to 200 feet thick.[13] The thicker seams in the West yield more coal per acre. Even though reclamation may be more difficult because a larger permanent amount of material has been removed, the reclamation cost can be averaged over greater tonnage. The draglines and power shovels in area mining can efficiently remove the overburden with relatively low cost, and much of the overburden can be deposited directly in the final desired location.

The costs in surface mining due to the state reclamation regulations were relatively small. A study by Herbert A. Howard indicated that reclamation requirements in eastern Kentucky could be met for less than $100 an acre.[14] S. Brock et al. computed the total costs for compliance with the 1967 West Virginia regulations to be $266 per acre.[15] The more recent study by Charles River Associates suggests a reasonable cost range for reclamation of area surface mines from $200 to 350 per acre and $500 to 750 per acre for contour mines.[16] It is likely that future reclamation requirements will attempt to restore surface-mined land to its "approximate original contour." The impending federal legislation will almost certainly contain this provision, and certain states in Appalachia have already adopted such legislation. Area mine operators in Kentucky, Maryland, Ohio, Pennsylvania, and West Virginia will be required to grade to the approximate original contour, but only Pennsylvania and Ohio will require the restoration of land to the approximate original condition, although terracing or other alternatives will be allowed under certain conditions.[17] How well these state regulations will be enforced, however, is still a question.[18] The state requirements of Kentucky, for example, appear to have been violated by more than one-third of the surface operations in eastern Kentucky. Part of the problem can occur since the land-reclamation bond per acre required to be filed as reported in most states is generally between $200 and $500, which is less than the actual costs of reclamation.

Surface production of acute slope angles in Appalachia, particularly Central Appalachia, will be affected most by "approximate original contour" requirements. In relatively flat or rolling topography, the operator of an area surface mine can generally restore the contour simply by filling and regrading the soil. In the West, the creation of a "rolling topography" will be required, probably with the stipulation that it be crossable by livestock or by agricultural machinery, depending on whether grazing or agriculture is planned. Western mines can be more than 100 feet deep, and strict adherence to "original contour" clauses would require that another large pit be dug somewhere to provide material to refill these sites—obviously an absurd interpretation of these laws. Western metals mining industries with large open-pit operations will also have to contend with future legislation, possibly with more problems than the coal industry since they have little experience with reclamation.

TABLE 6.3

Estimated Incremental Production Costs for Various Reclamation Costs

Region	Calculated Production per Acre Mined[a]	Reclamation Costs per Mined Acre, cents per ton			
		$1,000	$2,000	$3,000	$4,000
Appalachia					
Alabama	4,030	24.8	49.6	74.4	99.2
Kentucky (eastern)	4,460	22.4	44.8	67.2	89.6
Ohio	5,330	18.8	17.6	56.4	35.2
Pennsylvania	4,610	21.8	43.6	65.4	87.2
Tennessee	4,180	24.0	48.0	72.0	96.0
Virginia	5,900	17.0	34.0	51.0	68.0
West Virginia	7,060	14.2	28.4	42.6	56.8
Average	5,080	20.4	40.8	61.2	81.6
Central					
Illinois	7,200	13.8	27.6	41.4	55.2
Indiana	6,620	15.0	30.9	45.0	60.0
Kentucky (western)	7,340	13.6	27.2	40.8	54.4
Average	7,050	14.2	28.4	42.6	56.8
Western					
Colorado	12,100	8.2	16.4	24.6	32.8
Montana[b]	66,100	1.6	3.2	4.8	6.4
Wyoming	66,100	1.6	3.2	4.8	6.4
Average	48,000	3.8	7.6	11.4	15.2

[a]Based on density of 1,440 tons of bituminous coal per acre-foot at 80 percent recovery.
[b]Montana entry changed to reflect mining of subbituminous coal in Powder River Basin.
Source: Council on Environmental Quality, *Coal Surface Mining and Reclamation* (Washington, D.C.: Government Printing Office, 1973), p. 28.

This reliance on legal statute to control the side effects of surface mining stems from the failure of common law to eliminate the external effects. Everett F. Goldberg et al., discuss several factors which limit the effectiveness of litigation in controlling environmental damages.[19] Many cases of damage lack a plaintiff. It was common practice at the turn of the century for land owners in Appalachia to sell their "right to complain" about land-use impacts to the coal companies. Damage can occur from mines which have been abandoned, and yet the courts "for the most part have proved unwilling to impose liability on persons other than the actual mine operators (owners of the surface, or lessors, or predecessors in title to the coal rights)."[20] Private litigation has failed to eliminate the environmental problems of surface mining for several reasons, such as the cumbersome time lag for suits, court expenses, the risk of litigation from the possible loss of a suit, and the problems arising from using a single suit to claim full damages from several impacts of unreclaimed mining.

The cost of achieving higher levels of reclamation, such as returning land to original contour, was studied by the Council on Environmental Quality.[21] It was assumed that reclamation was carried out concurrently with the surface mining, rather than after mining activity had ceased, as in orphan mines. To estimate the average increase in production costs, reclamation costs were calculated in dollars per acre as incremental production costs in cents per ton. These are shown in Table 6.3. The cost of completely reclaiming contour mines in Appalachia was estimated to run up to 95 cents a ton. The expenditures at $4,000 per acre represent the Council's estimations for proposed reclamation, and we will use these figures in the discussion following. The alternative levels simulate expenditures if new techniques, mining methods, etc., were to lower these costs. The differences in regional mining conditions we have noted are reflected in the variability of the estimates. As expected, reclamation costs are highest in the Eastern fields and lowest in the West. The effects of imposing these reclamation costs will now be considered. The objective function (3.4) is modified to:

$$\text{Minimize } \Sigma \Sigma \Sigma \Sigma (C_{ij} + t_{il})X_{ijkl} + \phi r_{ij}X_{ijkl}$$
$$i \ j \ k \ l$$

where r_{ij} is the incremental production cost per ton for an expenditure of $1,000 per acre reclaimed in the ith supply district and ϕ is a policy parameter which allows us to simulate any desired level. Since the 1969 costs of reclamation are already approximated in our cost estimates, the CEQ estimates are introduced as increments.

Table 6.4 shows the effects of the imposition of reclamation policy on surface mining in relation to an absence of reclamation regulations in the basic model. The effects of the higher reclamation figures are seen in Table 6.5. One widely discussed hypothesis is that the share of underground mining of the total

TABLE 6.4

Regional Shipments with Surface-Mine Reclamation Costs by Mining Method
(millions of tons)

Region	Reclamation Costs per Acre			
	$1,000	$2,000	$3,000	$4,000
Northern Appalachia				
Surface	95.8	95.8	90.4	90.4
Underground	69.9	69.9	69.9	69.9
Total	165.7	165.7	160.3	160.3
Central Appalachia				
Surface	63.0	63.0	60.1	58.1
Underground	59.1	60.6	70.9	72.8
Total	122.1	123.6	131.0	130.9
Southern Appalachia				
Surface	14.6	14.6	14.6	14.6
Underground	5.5	5.5	5.5	5.5
Total	20.1	20.1	20.1	20.1
Appalachia, total				
Surface	173.4	173.4	165.2	163.2
Underground	134.5	136.1	146.3	148.3
Total	307.9	309.5	311.5	311.5
Midwest				
Surface	130.4	128.6	125.6	103.7
Underground	23.6	23.6	23.6	44.9
Total	154.0	152.2	149.2	148.6
West				
Surface	38.6	38.6	38.6	38.6
Underground	3.3	3.3	3.3	3.3
Total	41.9	41.9	41.9	41.9
United States				
Surface	342.3	340.6	329.4	305.4
Underground	161.4	163.0	173.3	196.5
Total	503.7	503.6	502.7	501.9

Source: Compiled by author.

TABLE 6.5

Comparative Regional Effects of the
Surface-Mine Reclamation Policy
(millions of tons)

Region	No Reclamation Policy	$4,000 Reclamation Cost per Acre	Net Change
Northern Appalachia			
Surface	95.8	90.4	−5.4
Underground	69.9	69.9	0.0
Total	165.7	160.3	−5.4
Central Appalachia			
Surface	63.0	58.1	−4.9
Underground	59.1	72.8	+13.7
Total	122.1	130.9	+8.8
Southern Appalachia			
Surface	14.6	14.6	0.0
Underground	5.5	5.5	0.0
Total	20.1	20.1	0.0
Appalachia, total			
Surface	173.4	163.1	−10.3
Underground	134.5	148.3	+13.8
Total	307.9	311.5	+3.5
Midwest			
Surface	130.4	103.7	−26.7
Underground	23.6	44.9	+21.3
Total	154.0	148.6	−5.4
West			
Surface	38.6	38.6	0.0
Underground	3.3	3.3	0.0
Total	41.9	41.9	0.0
United States			
Surface	342.4	305.4	−37.0
Underground	161.4	196.5	+35.1
Total	503.8	501.9	−1.9

Source: Compiled by author.

TABLE 6.6

Regional Shipments with Surface-Mine Reclamation Costs and Sulfur Emissions Standard
(millions of tons)

| Region | Reclamation Costs per Acre | | | |
	$1,000	$2,000	$3,000	$4,000
Northern Appalachia				
Surface	87.2	87.3	87.3	87.3
Underground	79.1	79.2	79.2	79.2
Total	166.3	166.5	166.5	166.5
Central Appalachia				
Surface	63.0	63.0	63.0	63.0
Underground	75.0	75.0	75.0	75.0
Total	138.0	138.0	138.0	138.0
Southern Appalachia				
Surface	14.6	14.6	14.6	14.6
Underground	5.5	5.5	5.5	5.5
Total	20.1	20.1	20.1	20.1
Appalachia, total				
Surface	164.8	164.9	164.9	164.9
Underground	159.7	159.7	159.7	159.7
Total	324.5	324.6	324.6	324.6
Midwest				
Surface	104.8	99.7	84.3	77.8
Underground	24.3	27.6	39.4	45.6
Total	129.1	127.3	123.7	123.4
West				
Surface	45.7	47.5	50.9	50.9
Underground	3.3	3.3	3.3	3.3
Total	49.0	50.8	54.2	54.2
United States				
Surface	315.3	321.1	300.2	293.6
Underground	187.3	190.7	202.4	208.6
Total	502.6	502.8	502.6	502.2

Source: Compiled by author.

TABLE 6.7

Comparative Regional Effects
of the Surface-Mine Reclamation Policy
Including the Sulfur Emissions Standard
(millions of tons)

	No Reclamation Policy	$4,000 Reclamation Cost per Acre	Net Change
Northern Appalachia			
Surface	92.7	87.3	−5.4
Underground	74.7	79.2	+4.5
Total	167.4	166.5	−0.9
Central Appalachia			
Surface	63.0	63.0	0.0
Underground	75.1	75.0	−0.1
Total	138.1	138.0	−0.1
Southern Appalachia			
Surface	14.6	14.6	0.0
Underground	5.5	5.5	0.0
Total	20.1	20.1	0.0
Appalachia, total			
Surface	170.3	164.9	−5.4
Underground	155.3	159.7	+4.4
Total	325.6	324.6	−1.0
Midwest			
Surface	120.1	77.8	−42.3
Underground	29.1	45.6	+16.5
Total	149.2	123.4	−25.8
West			
Surface	45.6	50.9	+5.3
Underground	3.4	3.3	−0.1
Total	49.0	54.2	+5.2
United States			
Surface	336.0	293.6	−42.4
Underground	187.8	208.6	+20.8
Total	523.8	502.2	−21.6

Source: Compiled by author.

production would increase. This might be expected since the relative cost differential between the mining methods has narrowed. Overall, the total tonnage of underground mining increases, with its share rising from 32 to 39 percent. The two underground areas mainly affected are Central Appalachia and the Midwest. The Midwest experiences both the largest increase in underground production (90 percent) and the largest decrease in surface mining. This occurs because the Midwest producers often serve many utility markets which are not in the producing area. Illinois producers, for example, serve Missouri, Tennessee, etc., as well as their own utilities. As the cost of surface reclamation reduces the per-ton extraction cost differential between surface and underground coal, the production of higher Btu underground coal in the Midwest becomes more competitive with surface mining. Surface shipments decrease in Central and Northern Appalachia, which face the highest reclamation costs in the East, and underground production increases in Central Appalachia. The West is unaffected by the policy.

Table 6.6 summarizes how these results would be affected if a sulfur emissions standard were added to the reclamation policy. Table 6.7 compares the effects of the reclamation policy at the highest CEQ estimates if the sulfur emissions standard is added to the model. There are some differences between Tables 6.5 and 6.7. Central Appalachia is now unaffected in this simulation since its low-sulfur coal, both surface and underground, is needed to meet the standard. We might expect that the Midwest would be affected most by these changes in surface mining. Not only does the sulfur standard result in a surface-mining decrease with a corresponding stimulation of underground production, as noted in the previous chapter, but the effects of the reclamation policy (in Table 6.5) accentuate these changes.

The interaction between these factors clearly is important, for underground production increases by 55 percent and surface-mining shipments fall drastically. Since lower-sulfur surface-mined coal is required, however, the increase in underground production is less than it would be without the emissions standard. Western surface activity increases to help replace Midwestern coals. Table 6.8 summarizes these effects by comparing the reclamation policy both with and without sulfur emissions regulation. The main conclusion that can be drawn is that Midwestern surface mining is affected by reclamation requirements far more with than without a sulfur standard. Central Appalachia, given its low sulfur production, is not substantially affected. Similarly, Midwestern underground production is increased more from the imposition of reclamation costs on local surface operations than from sulfur restrictions. Lower-sulfur underground production in the Midwest was increased by 5.6 million tons when the sulfur emissions standard was introduced, and it increases 16.5 million tons above this initial change after reclamation costs are added.

The changes in the delivered cost of coal to utilities are relatively minor. Consider a coal with 10,000 Btu per pound, for the sake of simplicity, which would have 20 million Btu per ton. A $1.00-per-ton increase in the production

TABLE 6.8

The Net Regional Effects of $4,000-per-Acre Reclamation Estimates with and without the Sulfur Emissions Standard (millions of tons)

Region	Without Standard	With Standard
Northern Appalachia		
Surface	−5.4	−5.4
Underground	0.0	+4.5
Total	−5.4	−0.9
Central Appalachia		
Surface	−4.9	0.0
Underground	+13.7	−0.1
Total	+8.8	−0.1
Southern Appalachia		
Surface	0.0	0.0
Underground	0.0	0.0
Total	0.0	0.0
Appalachia, total		
Surface	−10.3	−5.4
Underground	+13.8	+4.4
Total	+3.5	−1.0
Midwest		
Surface	−26.7	−42.3
Underground	+21.3	+16.5
Total	−5.4	−25.8
West		
Surface	0.0	+5.3
Underground	0.0	−0.1
Total	0.0	+5.2
United States		
Surface	−37.0	−42.2
Underground	+35.1	+20.8
Total	−1.9	−21.6

Source: Compiled by author.

TABLE 6.9

Acreage Disturbed and Reclaimed with and without a 2.00-Pound Sulfur Emissions Standard

Region	Reclamation Costs per Acre			
	$1,000	$2,000	$3,000	$4,000
Northern Appalachia				
Without standard	18.39	18.39	17.24	17.24
With standard	16.79	16.82	16.82	16.82
Central Appalachia				
Without standard	14.99	14.99	14.31	13.83
With standard	14.99	14.99	14.99	14.99
Southern Appalachia				
Without standard	3.62	3.62	3.62	3.62
With standard	3.62	3.62	3.62	3.62
Appalachia				
Without standard	37.00	37.00	35.17	34.67
With standard	35.40	35.43	35.43	35.43
Midwest				
Without standard	18.46	18.19	17.78	14.74
With standard	14.77	14.08	11.98	11.09
West				
Without standard	1.51	1.51	1.51	1.51
With standard	1.62	1.64	1.70	1.70
United States				
Without standard	56.97	56.70	54.46	50.94
With standard	51.79	51.15	49.11	48.22

Source: Compiled by author.

cost of surface-mined coal would result in an increase of only 5 cents per million Btu. Since most Eastern and Midwestern fields have values ranging from 11,000 to 12,000 Btu per pound, the average increase would be only around 4 cents from imposing reclamation. The largest increases occur in the Eastern markets, since the districts supplying these markets have the highest reclamation costs. The changes in the Western states are negligible, around 5 cents.

Tables 6.9 and 6.10 show the regional reclamation expenditures and land utilized in surface production at alternative reclamation estimates. The total acreage disturbed and reclaimed varies inversely with successive levels of reclamation. The relative share of underground production in total mining activity

TABLE 6.10

Total Reclamation Expenditures with and without a 2.00-Pound Sulfur Emissions Standard

| Region | Reclamation Costs per Acre | | | |
	$1,000	$2,000	$3,000	$4,000
Northern Appalachia				
Without standard	1,936,324	3,000,808	5,460,978	5,537,624
With standard	1,776,571	2,863,238	5,337,328	5,726,477
Central Appalachia				
Without standard	1,533,125	3,000,954	4,295,655	5,536,116
With standard	1,533,125	3,000,954	4,501,431	6,001,908
Southern Appalachia				
Without standard	362,204	724,408	1,086,612	1,448,816
With standard	362,204	724,408	1,086,612	1,448,816
Total, Appalachia				
Without standard	3,831,653	6,726,170	10,843,246	12,522,557
With standard	3,671,900	6,588,601	10,925,371	13,177,202
Midwest				
Without standard	1,837,412	3,653,347	5,310,429	5,866,200
With standard	1,471,798	2,825,998	3,580,666	4,418,344
West				
Without standard	152,551	305,103	537,334	610,207
With standard	163,987	333,497	516,652	688,869
United States				
Without standard	5,821,617	10,684,621	16,691,010	18,998,965
With standard	5,307,685	9,748,097	15,022,690	18,284,416

Source: Compiled by author.

increases as required reclamation expenditures increase. The total surface-mine expenditures for reclamation per year increase as we impose these successive reclamation costs. These total costs of reclamation reflect the annual resource expenditure in capital equipment, wages, etc., needed at each of the reclamation levels. Reclamation expenditures at the $4,000-per-acre level are over three times as large as those at the $1,000-per-acre level, even though relative land utilization in surface-mining activities is only 89 percent at the higher reclamation estimates.

The study by the National Academy of Sciences has estimated that reclamation would add but a few cents per ton to the productions costs of coal in the main Western fields.[22] Since the CEQ study based its estimates for Montana and

Wyoming on the relatively thick seam conditions in the Powder River Basin, we might wish to consider whether higher reclamation values in the West would influence our results. The highest published estimate for reclamation in these areas is $4,000-5,600 per acre, cited by the Peabody Coal Company for its Montana operation.[23]

Assuming a low coal-seam thickness of only 10 feet, additional production costs would be a maximum of only 28 cents a ton.[24] If we impose these higher costs on the general Northern Great Plains area in the model, what is affected? Practically nothing. The explanation is quite simple, for the additional increases would add only around 1.2 cents per million Btu to the cost of coal from these areas. This value is too low to affect either Western shipments to the Midwest or the pattern of intra-Western deliveries. This situation exists even if we consider the imposition of a strict reclamation policy on the Western producers alone.

Opponents of surface mining in the West, seeing the environmental degradation in the more established mining areas, were afraid that the coal industry would "Appalachianize" their region.[25] As Senator Mike Mansfield of Montana put it, the West could become the "utility for the rest of the nation."[26] It seemed that Western producers would be forced to undertake high levels of reclamation, even if the contour surface mines of Appalachia were not. Somewhat ironically, the largest percentage of surface-mined land which state legislation will require a return to "approximate original contour" will be in the Appalachian area. not the West.[27] Again, deliveries from the West are not affected by these reclamation costs.

These factors, as well as those discussed earlier, lead us to a general conclusion concerning the effects of a strict reclamation policy on the Western coal industry. The use of Western coal is relatively unaffected by additional production costs incurred from reclamation. The crucial determinant for the use of Western coal in non-Western markets is its relatively low sulfur level. If we impose a policy which regulates sulfur emissions, Western coal will be used in the Midwest with the local high-sulfur coal.

The cost of reclamation for contour surface mines in Appalachia, particularly as an increasing function of slope angle, has been an issue of debate centering around the relationship between earth-moving costs and the method of surface mining (as opposed to revegetation cost components, etc.). In conventional contour strip mining, the overburden is simply thrown down the slope and stacked into a large mound. Reclamation to "approximate original contour" entails bringing this material back up the slope, stacking it, and finally regrading to the original contour.

The costs of this type of contour surface-mine reclamation, known as "backfilling," have been extensively studied as part of the Appalachian Resources Project (ARP) at the University of Tennessee.[28] It is concluded that the CEQ figures on backfilling costs are too low for Appalachia by a factor of approximately two.[29] Under similar assumptions as the CEQ study, the ARP finds reclamation costs as a function of slope angle at 91 cents per ton at 20°, $1.34 at 25°, $1.94 at 30°, and $2.65 at 35°.[30] These figures represent maximums since they are based on conventional backfilling techniques. Revegetation costs generally are 20 cents per ton of these figures. If the present work of Consolidation Coal Company is successful in growing grass annually for grazing on acid refuse without the use of topsoil, these revegetation costs would be drastically reduced.

Other mining techniques include reclamation as an integral part of the mining operation. In the modified block-cut technique, for example, each area which has been mined is filled with the overburden from the succeeding surface cuts. Furthermore, (except for the first cut) the spoil is not pushed down the slopes and almost all of the disturbed acreage is restored to the original contour immediately. This significantly reduces earth-moving costs below those of conventional backfilling. The current experience in Pennsylvania, as well as new data developed by the ARP research, indicates that using the modified block-cut technique on steep slopes can lower reclamation costs to near $1.00 per ton, which is the general figure used by the CEQ study.[31] Experience with reclamation in general, as well as the development of new reclamation techniques, has made Appalachian coal operators more optimistic about the impact of legislation on their operations. As Dan Gerkin of the West Virginia Surface Mining and Reclamation Association put it, "H.R. 1500 (original contour bill) won't have great impact here."[32]

The ARP research indicates that an average figure for the effects of conventional contour with backfilling reclamation would be $2.00 per ton.[33] Hubert Hagen has suggested that the average figure for area mine reclamation in the Midwest would be approximately $1.34 per ton.[34] These suggestions could add a $1.00 per ton to the CEQ study estimates. These levels of reclamation again would not have a large effect on the delivered costs of coal. In the East, the overall increase per million Btu would be only 8 cents, and in the Midwest around 6 cents per million Btu. This increase in production costs in the Midwest would not create additional demand for Western coal in this market beyond that created by the sulfur emissions standard.* Since we were interested in the degree to which the relative position of underground and surface production would be altered in the Midwest and Appalachia, we intended to use the higher estimates. However, it now appears that additional reclamation, while increasing the delivered price of coal per million Btu somewhat, would have no real effect. The effect is minimized by a change in the institutional structure of underground mining which has not been widely publicized outside the industry itself.[35]

---

*The use of Western coal averages sulfur emissions to meet the standard; it is never used from a relative cost advantage alone.

The coal industry, given a two-year extension from the original 1975 implementation date, must take over the payments of the black lung compensation program from the federal government by 1977, roughly at the same time that reclamation statutes will begin to be enforced. The payments of smaller mine operators could actually exceed their current payroll.[36] As a result, underground operators in the East and Midwest have begun to contract for insurance against black lung disease and subsequent payments to workers.* The industry costs of the insurance programs will be approximately $1.00 per underground ton. By the structure of linear programming models, adding this same figure to both surface and underground mining in Appalachia and the Midwest will not alter the regional pattern of distribution and utilization except for the effects of an increase in coal costs of approximately 4 cents per million Btu.[37]

The expectation that the higher reclamation increments would force the coal industry to turn to underground mining in these areas has been offset further by a change in the institutional structure of the industry. The black-lung compensation plan encompasses the theory of "last responsible operator," which may have serious consequences for the future employment patterns of underground miners. Under this Department of Labor regulation, an operator who employs a miner for one year is responsible for that worker's black-lung claims, even though symptoms can take time to appear and the disease could have been caused by years of exposure at other mines. Mine operators, therefore, might be reluctant to hire experienced miners, even though their skills are at a premium today, rather than risk the responsibility for their long-term disability payments. Under the insurance programs, this employment rigidity has been decreased.

THE IMPACT OF A BAN ON HIGH-SLOPE SURFACE-MINING PRODUCTION

We have already considered the problems which can arise from contour surface mining on steep slopes and the difficulty in maintaining soil stability even after reclamation. There are natural limits to the slope angle that will support loose spoil without reclamation and even after reclamation in the high rainfall of Appalachia. The critical grade at which soil is stabilized is not an absolute one because of the wide variability of site-specific mining situations. As early as 1967 a slope angle of 20° was suggested as the critical level since stability above this angle is difficult to achieve.[38]

---

*TVA keeps one of the most consistent current recording systems for the cost components from its coal suppliers, both large and small. We thank Lee Sheppard, Office of Information, for providing these data and information. The Western underground mines, probably since they are generally the newest mines, appear not to have contracted for insurance to the best of our knowledge.

TABLE 6.11

Percentage Changes in Coal Extraction
for a 20°-Slope-Angle Surface-Mine Ban in 1971

Region	CEQ Scenario		
	High Impact	Medium Impact	Low Impact
Northern Appalachia	−12.0	−11.8	0.0
Central Appalachia	−33.5	−21.9	−10.2
Southern Appalachia	−17.6	0.0	0.0
Appalachia, total	−22.5	−11.3	−4.9

Source: Council on Environmental Quality, *Coal Surface Mining and Reclamation* (Washington, D.C.: Government Printing Office, 1973), pp. 51, 66.

As a result of congressional interest in surface-mining legislation, the CEQ considered the impact of a ban on coal surface mining on high slopes, primarily those greater than 20°.[39] The adverse economic impact of three different levels in Appalachia in 1971 were considered. Each of the three levels was determined from an assumed increase in coal extracted both from lower-slope surface mines and from underground mines in Appalachia as a result of the ban.[40] The greatest impact resulted from the assumption that no losses in Appalachian coal production would be compensated for by other mining activities in Appalachia. These losses in production would presumably be made up by increases in coal extraction outside of Appalachia and increases in the utilization of other fossil fuels. No attempt was made in the CEQ study to estimate the impact of the ban on the other coal producing areas or the implications for natural gas and oil. A low-impact scenario assumed the best possible response from other production in Appalachia. All losses in northern and southern Appalachia would be made up by increases in alternative sources from these same regions. Central Appalachian losses would be made up, at least partially, by a 10-percent increase in underground production in the area. The third scenario considered a medium impact, where the assumptions were generally the same as those in the second case, except that losses in Central Appalachia were compensated for by only a 5-percent increase in local underground production.

The economic impact measured in the CEQ study included changes in production for each of the three scenarios. Table 6.11 gives a summary of the net yearly losses in regional coal extraction from a ban in 1971 on coal surface mining on slopes exceeding 20°, as estimated in the study. For purposes of comparison with our results, these estimates are presented in terms of percentage changes.

The effects of the 20° slope angle limitation in our basic model are given in Table 6.12. As expected, Central Appalachia suffers the largest decline in

TABLE 6.12

Comparative Regional Effects of the 20°-Slope-Angle Surface-Mine Ban by Mining Method
(millions of tons)

Region	Without Ban	With Ban	Net Change
Northern Appalachia			
Surface	95.8	66.1	−29.7
Underground	69.9	81.1	+11.2
Total	165.7	147.2	−18.5
Central Appalachia			
Surface	63.0	17.2	−45.8
Underground	59.1	75.0	+15.9
Total	122.1	92.2	−29.9
Southern Appalachia			
Surface	14.6	14.6	0.0
Underground	5.5	20.1	+14.6
Total	20.1	34.7	+14.6
Total, Appalachia			
Surface	173.4	97.8	−75.5
Underground	134.5	176.5	+41.7
Total	307.9	274.0	−33.8
Midwest			
Surface	130.4	130.6	+0.2
Underground	23.6	61.8	+38.2
Total	154.0	192.4	+38.4
West			
Surface	38.6	53.9	+15.3
Underground	3.3	4.1	+0.8
Total	41.9	58.0	+16.1
United States			
Surface	342.4	282.3	−60.1
Underground	161.4	242.1	+80.7
Total	503.8	524.4	+20.6

Source: Compiled by the author.

TABLE 6.13

Changes in the Delivered Cost of Steam Electric Coal Due to the 20°-Slope-Angle Ban on Surface Mining

State	Without Ban	With Ban	Net Change
Alabama	0.49	0.63	0.14
Arizona	0.36	0.36	0.00
Arkansas	—*	—	—
California	—	—	—
Colorado	0.40	0.42	0.02
Connecticut	0.60	0.78	0.18
Delaware	0.50	0.67	0.17
District of Columbia	0.52	0.69	0.17
Florida	0.60	0.74	0.14
Georgia	0.49	0.64	0.15
Idaho	—	—	—
Illinois	0.45	0.49	0.04
Indiana	0.40	0.46	0.06
Iowa	0.50	0.55	0.05
Kansas	0.50	0.54	0.04
Kentucky	0.39	0.53	0.14
Louisiana	—	—	—
Maine	—	—	—
Maryland	0.52	0.69	0.14
Massachusetts	0.63	0.81	0.18
Michigan	0.54	0.72	0.18
Minnesota	0.59	0.59	0.00
Mississippi	0.54	0.62	0.08
Missouri	0.42	0.46	0.04
Montana	0.31	0.31	0.00
Nebraska	0.49	0.51	0.02
Nevada	0.52	0.52	0.00
New Hampshire	0.64	0.82	0.18
New Jersey	0.57	0.74	0.17
New Mexico	0.30	0.30	0.00
New York	0.58	0.75	0.17
North Carolina	0.49	0.67	0.18
North Dakota	0.40	0.40	0.00
Ohio	0.48	0.67	0.19
Oklahoma	0.49	0.52	0.03
Oregon	—	—	—
Pennsylvania	0.44	0.62	0.18
Rhode Island	0.63	0.81	0.18
South Carolina	0.53	0.71	0.18
South Dakota	0.53	0.53	0.00
Tennessee	0.39	0.57	0.18
Texas	0.54	0.54	0.00
Utah	0.39	0.39	0.00
Vermont	0.63	0.81	0.18
Virginia	0.52	0.70	0.18
Washington	—	—	—
West Virginia	0.41	0.59	0.18
Wisconsin	0.52	0.70	0.05
Wyoming	0.31	0.32	0.01

*No demand for steam coal by utilities in the state.
Source: Compiled by author.

surface production, almost 46 million tons, since high-slope mining is concentrated in that area. This decline is offset partially by a 23-percent increase in Appalachian underground production, which helps to limit the total decline in Appalachian production. However, production from outside the Appalachian area, particularly underground activity in the Midwest and surface production in the West, also compensates for losses. The overall increase in underground production of 81 million tons is divided almost equally between the Midwest and Appalachia. Surface production in the West significantly increases as a result of the ban.

The ban increases the total emissions of sulfur by 8 percent, or approximately 0.9 million tons. This represents an increase in sulfur oxide pollutants of almost 1.8 million tons.

The costs of the ban to utilities on a regional basis are given in Table 6.13. As we might expect, the largest increases occur in those markets traditionally served by the surface production of Appalachia. These areas must now use more distant supply sources, with subsequently higher delivered costs. The largest increases occur in the Eastern states and coal-producing states affected by the ban. The general increase in delivered costs in such states as New York, Pennsylvania, and Ohio is 18 cents per million Btu. The increase in costs in the Midwest of 4 or 5 cents per million Btu generally represents the increased reliance on local underground coal with its higher extraction costs. Not too surprisingly, the Western states experience no change. Overall, the increases in costs do not seem to be too severe.

The main conclusion to be drawn from an analysis of the ban is that it would serve as a major impetus to local underground production in Appalachia and as an incentive for increased activity by Midwestern underground producers to capture partly what were previously "Appalachian" markets. Underground production in the West is significantly increased, by 15.2 million tons or approximately 39.4 percent. The degree to which the reader would classify the net environmental impact in this case depends on a personal evaluation of the diverse regional effects. The problems of high-slope surface production in Appalachia are decreased, and underground production is stimulated. Problems of adequate surface-mine reclamation are intensified in the West. Since our analysis has indicated that the ban would stimulate substantial increases in underground mining, the net environmental effect could be positive. We assume two points implicitly here, however. The first is that adequate reclamation in the West is indeed feasible. Second, the human costs of a switch to underground mining from surface production are not severe.

One of the most important facts developed from the data of the CEQ study was that the sulfur content of Central Appalachian coal deposits is inversely correlated to increasing slope angle. Table 6.14 shows this relationship. In 1971, 23 percent of all low-sulfur coal (less than 1 percent) used in conventional steam electric generation was extracted from high-slope surface mines in Central

TABLE 6.14

Sulfur Content of Central Appalachian Surface-Mined Coal as a Function of Slope Angle
(millions of tons per year)

Sulfur Content / Percent	Total	0-9.9°	10-14.9°	15-19.9°	20-24.9°	25°+
Slope Angle						
1	40.17	0.35	0.53	0.75	6.58	31.95
1-1.49	15.80	0.12	0.41	0.44	3.48	11.35
1.5-1.99	7.30	0.09	0.37	4.12	0.95	1.77
2-2.99	1.42	0.19	0.32	0.22	0.36	0.33
3	0.34	0.00	0.00	0.34	0.00	0.00
Total	65.03	0.75	1.63	5.87	11.37	45.41
Percentage of Total Production						
1	61.8	0.5	0.8	1.1	10.1	49.1
1-1.49	24.3	0.2	0.6	0.7	5.4	17.5
1.5-1.99	11.2	0.1	0.6	6.3	1.5	2.7
2-2.99	2.2	0.0	0.0	0.5	0.0	0.5
3	0.5	0.0	0.0	0.5	0.0	0.5
Total	100.0	1.1	2.5	8.9	17.6	69.8

Source: Based on United States Bureau of Mines Field Survey, January, 1973. See Council on Environmental Quality, *Coal Surface Mining and Reclamation* (Washington, D.C.: Government Printing Office, 1973), Appendix G, and p. 54.

TABLE 6.15

Strippable Reserves in Appalachia as a Function of Slope Angle
(millions of tons)

| State | Total | Slope Angle | | | | | Total Deep Reserves in Appalachia |
		0-9.9°	10-14.9°	15-19.9°	20-24.9°	25°+	
Alabama	169.84	124.79	16.41	13.19	10.04	5.40	12,774
Kentucky	766.52	44.80	38.84	106.36	219.36	357.36	37,639
Maryland	27.27	25.17	1.71	0.26	0.13	0.00	1,117
Ohio	1,334.01	961.04	256.44	101.92	13.42	0.00	36,505
Pennsylvania	1,293.48	1,116.24	161.34	10.16	3.42	2.45	66,011
Tennessee	135.66	75.85	8.51	22.24	24.26	4.80	2,094
Virginia	226.86	0.00	0.00	32.06	131.78	63.02	8,324
West Virginia	2,507.01	364.52	592.04	475.87	608.86	465.72	90,059
Total	6,460.65	2,712.41	1,075.30	763.06	1,011.27	898.55	254,523
Percentage	100.00	42.00	16.60	11.80	15.70	13.90	

Source: All data sources and analytical techniques are described in Council on Environmental Quality, *Coal Surface Mining and Reclamation* (Washington, D.C.: Government Printing Office, 1973), Appendix G.

TABLE 6.16

Comparative Regional Effects of the 20°-Slope-Angle Surface-Mine Ban by Mining Method with the 2.00-pound Sulfur Emissions Standard (millions of tons)

Region	Without Ban	With Ban	Net Change
Northern Appalachia			
Surface	92.7	66.0	−26.7
Underground	74.7	82.8	+11.2
Total	176.4	148.8	−18.5
Central Appalachia			
Surface	63.0	17.2	−45.8
Underground	75.1	75.1	0.0
Total	138.1	92.3	−45.8
Southern Appalachia			
Surface	14.6	12.8	−1.8
Underground	5.5	5.5	−1.8
Total	20.1	18.3	−1.8
Total Appalachia			
Surface	170.3	96.0	−74.3
Underground	155.3	163.4	+8.1
Total	325.6	259.4	−66.2
Midwest			
Surface	120.1	136.4	+16.3
Underground	29.1	37.3	+8.2
Total	149.2	173.7	+24.5
West			
Surface	45.6	99.3	+53.7
Underground	3.4	6.1	+2.7
Total	49.0	105.4	+56.4
United States			
Surface	336.0	331.7	−4.3
Underground	187.8	206.4	+28.6
Total	523.8	538.1	+14.3

Source: Compiled by author.

Appalachia.[41] If production losses from the ban were replaced by higher-sulfur Midwestern coals, the CEQ estimates that sulfur oxide emissions would increase between 0.43 and 1.5 million tons, representing an increase in sulfur emissions of approximately 0.22 to 0.75 million tons.[42]

The concern over sulfur emissions has resulted in an extensive effort to mine the lower-sulfur coals in Central Appalachia, which are deposited on the steeper slopes. Table 6.15 clearly points this out. Although strippable reserves, as a function of slope angle, are concentrated on the lower slopes, current production is intensive on slopes above 20°. Total strippable reserves above 20° slopes equal 29.6 percent of Appalachia's reserves, yet 51.4 percent of current production come from these slopes. This pattern can be explained by the data in Table 6.14, showing that these higher slopes contain the lower-sulfur deposits. Therefore, the attempt to improve air quality by using lower-sulfur coals has aggravated the land-use impact of surface mining in locations where these effects are most severe.

The results of the ban in the model, in conjunction with a sulfur emissions standard, are shown in Table 6.16. As shown, substitution of production from lower-slope surface mines and underground mines in Appalachia does not significantly reduce the need for additional production from other areas. The decline in Appalachian surface mining is made up mainly by an additional increase in surface-mined production from the Midwest and West of 16.3 and 53.7 million tons, respectively. Total production increases by 14.3 million tons, since the Btu values of the lost Appalachian production, particularly of Central Appalachia, are among the highest of U.S. coal deposits. This production is replaced by other surface-mined coal that has a lower heating value, and subsequently additional production is required.

The changes in the delivered costs of utilities coal are given in Table 6.17. The increases for steam coal are generally quite large. Given our results, we would expect this. Most of the compensation for the decline in Appalachian production is an increase in shipments of Western coal. Even though the surface extraction costs per ton in the West are the lowest in the United States, the prices for this coal delivered in distant major markets are quite high. Not too surprisingly, the greatest increases occur in the Eastern markets, generally 60 cents per million Btu. There are similar large increases in costs in the Midwest. The smallest cost increases occur in the Western states and certain areas of the upper Midwest.

There are a few anomalies in the results. North Carolina has an increase of 109 percent in steam coal costs while South Carolina has only a 48-percent increase. Also, Wyoming has a cost increase twice that of Montana. These cost increases are higher when the 20°-slope-angle ban and the sulfur emissions standard are both in effect. We have noted earlier than an increase of 29 percent in coal cost to utilities would raise utilities costs by about 5.8 percent. For

TABLE 6.17

Changes in the Delivered Cost of Steam Coal
Caused by the 20°-Slope-Angle Ban on Surface Mining

State	Price of Steam Coal, dollars per million Btu		
	Without Ban	With Ban	Net Change
Alabama	0.49	1.03	0.54
Arizona	0.36	0.60	0.24
Arkansas	—*	—	—
California	—	—	—
Colorado	0.40	0.76	0.36
Connecticut	0.64	1.26	0.62
Delaware	0.52	1.11	0.59
District of Columbia	0.55	1.13	0.58
Florida	0.64	1.23	0.59
Georgia	0.53	1.14	0.61
Idaho	—	—	—
Illinois	0.68	1.13	0.45
Indiana	0.61	1.17	0.56
Iowa	0.55	1.01	0.46
Kansas	0.55	0.96	0.41
Kentucky	0.50	1.10	0.60
Louisiana	—	—	—
Maine	—	—	—
Maryland	0.56	1.16	0.60
Massachusetts	0.67	1.29	0.62
Michigan	0.59	1.16	0.57
Minnesota	0.59	0.87	0.28
Mississippi	0.61	1.02	0.41
Missouri	0.66	1.11	0.45
Montana	0.31	0.48	0.17
Nebraska	0.49	0.94	0.45
Nevada	0.52	0.78	0.26
New Hampshire	0.68	1.30	0.62
New Jersey	0.60	1.22	0.62
New Mexico	0.30	0.52	0.22
New York	0.61	1.23	0.62
North Carolina	0.52	0.90	0.38
North Dakota	0.40	0.56	0.16
Ohio	0.59	1.16	0.57
Oklahoma	—	—	—
Oregon	—	—	—
Pennsylvania	0.47	1.05	0.58
Rhode Island	—	—	—
South Carolina	0.56	1.17	0.61
South Dakota	0.53	0.69	0.16
Tennessee	0.42	1.03	0.61
Texas	—	—	—
Utah	0.39	0.67	0.28
Vermont	0.67	1.29	0.62
Virginia	0.55	0.93	0.38
Washington	—	—	—
West Virginia	0.44	1.03	0.59
Wisconsin	0.69	1.09	0.40
Wyoming	0.31	0.67	0.36

*No steam coal demand by electric utilities in state.
Source: Compiled by author.

example, an increase of 60 cents per million Btu for coal-fired generation would therefore raise total utility costs by almost 34.8 percent.

The interaction between concern for air quality and land-use impact can be seen by comparing the alternative results of the model in response to the 20°-slope-angle ban, depending on whether or not an explicit sulfur emissions standard is in force. When sulfur emissions were not a policy concern, Appalachian underground production increased by 41.7 million tons and represented the largest mining activity to compensate for the high-slope production losses. Much of this additional production was also lower-sulfur coal, particularly the increases in Central and Northern Appalachia. As we saw in the preceding chapter, this underground coal was necessary to meet a sulfur emissions standard even without the 20°-slope-angle ban and, therefore, cannot be a source of compensation for further surface losses. Underground production in Southern Appalachia, generally 2-3 percent sulfur coal, which increased 15 million tons previously, can no longer serve as compensation. Similarly, the changes in underground production, particularly from the Midwest, are now moderated by the higher sulfur levels of its coal deposits. Almost 19 million tons from the Midwest, which were used under the ban without the sulfur emission standard, are no longer allowable under the standard. Western production, which doubled with the introduction of the standard, increases mainly due to additional consequences of the sulfur emissions regulation. The West increased its surface production by an additional 46 million tons, a bigger increase than when we consider only the slope-angle prohibitions.

Table 6.18 summarizes the regional percentage changes in coal extraction caused by the ban on coal surface mining on slopes above 20°. The largest effect is clearly the difference caused in the West by the presence of the sulfur emissions

TABLE 6.18

Percentage Changes in Coal Extraction for a 20°-Slope-Angle Ban

Region	Without Sulfur Standard	With Sulfur Standard
Northern Appalachia	−11.2	−11.2
Central Appalachia	−24.6	−33.2
Southern Appalachia	+25.0	−9.0
Appalachia, total	−11.0	−20.3
Midwest	+25.0	+16.4
West	+38.3	+115.1
United States, total	+4.1	+2.7

Source: Compiled by author.

TABLE 6.19

Distribution of Steam Coal Extraction
with the 20°-Slope-Angle Ban

Region	Without Sulfur Standard		With Sulfur Standard	
	millions of tons	percent	millions of tons	percent
Appalachia	274.1	52.3	259.4	50.0
Midwest	192.4	36.7	173.7	30.5
West	58.0	11.0	105.4	19.5
United States	524.5	100.0	538.1	100.0

Source: Compiled by author.

standard. By comparing our results with the three scenarios developed by the CEQ in Table 6.11, we can see that the presence of the sulfur emissions standard is quite important. If we do not have a policy that attempts to regulate sulfur emissions, we see that the changes in our model suggest that the "medium-impact" scenario of the CEQ study is most likely to occur. With the sulfur emissions standard, the "high-impact" scenario would probably occur in Appalachia. The regional distribution of total production with the emissions standard indicates, as we might expect, a greater concentration in the Western producing areas and a decrease in the use of Midwestern coal, as seen in Table 6.19.

Total sulfur emissions are significantly decreased if the standard is present. Without the standard, sulfur emissions are decreased by 10.4 percent, or 1.15 million tons of emitted sulfur. In comparison to the results of the emissions standard policy without the 20°-slope-angle ban, however, total emissions do increase by 4.7 percent, since the availability of the lowest-sulfur coals is decreased by the ban.

The main conclusion from our analysis with the sulfur emissions standard is that the ban on high-slope surface mining would stimulate substantial increases in surface mining in the Midwest and West. During the operational time period of our model, there would be little compensation in underground and lower-slope surface production in Appalachia. The net improvement in environmental quality from the ban is problematic. The problems of high-slope contour surface mining in Appalachia would be minimized, of course, but the ban would shift the location of surface mining regionally to other areas. Unlike the results of our model, when no emissions standard is in effect the substantial increases occur in surface mining with only small additional increases in underground production. Given the ambiguity of the environmental effects of the ban and the

TABLE 6.20

Surface-Mine Employment in Appalachia as Function of Slope Angle, 1970
(number of miners)

Economic Area	Surface-Mine Employment	Slope Angle					Underground-Mine Employment
		0-9.9°	10-14.9°	15-19.9°	20-24.9°	25°+	
Williamsport, Pa.	1,483	548	445	356	104	30	661
Pittsburgh, Pa.	4,837	1,505	1,765	838	335	394	23,348
Cleveland, Ohio	1,099	112	0	70	693	224	212
Columbus, Ohio	1,091	70	258	763	0	0	491
Clarksburg, W.Va.	1,527	0	0	76	1,451	0	8,020
Huntington, W.Va.– Ashland, Ohio	5,027	127	202	211	532	3,955	30,699
Lexington, Ky.	2,080	0	0	0	790	1,290	5,025
Bristol, Va.	1,488	0	11	27	125	1,325	16,757
Knoxville, Tenn.	1,614	13	38	745	378	440	3,490
Nashville, Tenn.	117	112	5	0	0	0	67
Chattanooga, Tenn.	145	61	20	23	23	18	350
Birmingham, Ala.	1,337	562	187	214	214	160	3,674
Total	21,845	3,110	2,931	3,323	4,645	7,830	92,714
Percent	100.0	14.2	13.4	15.2	21.3	35.8	100.0

Source: Council on Environmental Quality, *Coal Surface Mining and Reclamation* (Washington, D.C.: Government Printing Office, 1973), p. 74.

large cost increases to Eastern coal users, the policy appears costly in relation to specific regional benefits. In our opinion, the 20°-slope ban appears to be another example of a policy in which a new environmental problem is created in the process of reducing existing environmental degradation.[43]

The long-term changes in Appalachian coal extraction in response to the slope-angle ban would depend on the extent to which coal from lower-slope surface mines and underground mines would be available to substitute for the losses from high slopes. We have seen that changes in lower-slope production, either from underground or surface mines, which would be required to be low-sulfur coal, would be relatively small. Although almost 59 percent of total strippable reserves are found on slopes less than 15°, their sulfur levels would discourage increases in new mine investment. In Central Appalachia, the substitution possibilities are further limited by the physical availability and slope location of reserves. There are few coal reserves on the lower slopes of eastern Kentucky and western Virginia of the magnitude of those lost by a ban. Even though the situation is not as severe for West Virginia and Tennessee, the reserves lost by a 20°-slope-ban are almost 45 percent of total strippable reserves in these areas.

Surface mine operators indicate that relocating equipment, establishing new facilities, etc., for a new operation in the same general area of an older operation can take up to eight months. Over the short run, any immediate increase in the capacity of underground mines would be slight. Existing deep mines may be able to increase production as much as 10 percent per year, but only if mining equipment is delivered on a continual basis.[44] Besides lags of up to several years for equipment delivery, rapid expansion in underground mining is further limited by long lead time required to open new mines and the necessity to insure returns on capital investments which can range from $8 to $20 per ton for utility-grade coal. The equipment used in surface mining, though efficient for its designed purpose, is not a substitute for specialized underground equipment. Furthermore, we have seen that the different skills required for surface and underground miners would inhibit a fast, smooth employment transition from surface to underground mining. It is apparent that neither displaced contour surface operations nor their employees could make the transition to underground mining "overnight."[45]

The distribution by slope angle of surface-mine employment in Appalachia is shown in Table 6.20. The approximate loss in surface-mine employment from a 20°-slope-angle-ban would be around 15,000 miners. Given the lower productivity of underground mining, an additional 42,200 underground miners might be needed to replace the lost surface production.[46]

The major response to the combined 20°-slope-angle ban and the sulfur emissions standard was an increase in surface production in regions outside of Appalachia. Western surface production climbed greatly, by 115 percent. As a result, strict reclamation would almost certainly be enforced in the Midwest and

TABLE 6.21

Comparative Regional Effects of the 20°-Slope-Angle Ban by Mining Method with Surface-Mine Reclamation Cost (millions of tons)

Region	Without Reclamation Policy and Ban	With $4,000-per-Acre Reclamation Cost and Ban	Net Change
Northern Appalachia			
Surface	92.7	66.1	−26.6
Underground	74.7	81.2	+6.5
Total	176.4	147.3	−20.1
Central Appalachia			
Surface	63.0	17.2	−45.8
Underground	75.1	75.0	+0.1
Total	138.1	92.2	−45.7
Southern Appalachia			
Surface	14.6	14.6	0.0
Underground	5.5	5.5	0.0
Total	20.1	20.1	0.0
Appalachia, total			
Surface	170.3	97.8	−72.5
Underground	155.3	161.7	+6.4
Total	325.6	259.5	−66.1
Midwest			
Surface	120.1	108.5	−11.6
Underground	29.1	45.6	+16.5
Total	149.2	154.1	+4.9
West			
Surface	45.6	99.5	+53.9
Underground	3.4	6.0	+2.6
Total	49.0	105.5	+56.5
United States			
Surface	336.0	305.8	−30.2
Underground	187.8	213.4	+25.6
Total	523.8	519.2	−4.6

Source: Compiled by author.

West. Therefore, we considered the effects of the ban in conjunction with reclamation requirements on surface-mined land unaffected by the slope limit. The effects of imposing the reclamation regulations of the model are given in Table 6.21. The increase in surface production in the West and Midwest is limited somewhat by the reclamation policy, since underground production in the Midwest is stimulated. Overall, the relative regional effects in this case are similar to those of the reclamation policies shown in Table 6.3. Midwestern underground production increases, as expected, and local surface production declines somewhat. The effects on the delivered costs of steam electric coal, sulfur emissions levels, etc., are negligible. Again this suggests, as with the transport rate simulation in the preceding chapter, that the effects of policy in the model are fairly stable.

The CEQ study first approximating the effects of a slope-angle ban concluded that "because the overwhelming majority of U.S. reserves are recoverable only by underground mining and because of large and as yet untapped reserves in the West, the loss of reserves from a slope-angle ban represents only about 1 percent of the total reserves."[47] In calling for slope-angle legislation, Rep. K. Hechler (West Virginia) considers this significant, for "we are only talking about a paltry 1 percent of the total reserves."[48] Our results would suggest that the location and sulfur content of these reserves and the associate costs to both utilities and other producing regions of a ban certainly make high-slope deposits an extremely important part of coal resources, particularly over the short run.

NOTES

1. This discussion is based on the recent United States Bureau of Mines study, *Land Utilization and Reclamation in the Mining Industry, 1930-1971*, IC8642 (Washington, D.C.: Government Printing Office, 1974). We thank Leonard Westerstrom of the Division of Fossil Fuels for making an early copy available to us.

2. Congressional Research Service, *Factors Affecting the Use of Coal in Present and Future Energy Markets* (Washington, D.C.: Government Printing Office, 1973), p. 25.

3. Ibid.

4. William Miernyk, "Environmental Management and Regional Economic Development" (paper presented at the Southern Economic Association meetings, Miami Beach, Florida, November 6, 1971).

5. Robert Spore, "The Economic Problem of Coal Surface Mining," *Environmental Affairs* 2, no. 4 (June 1973): 585-93.

6. A publication which emphasizes these activities is Mined Land Conservation Conference, *What About Strip Mining* (Washington, D.C.: Mined Land Conservation Committee, 1964).

7. F. K. Schmidt-Bleek et al., *Benefit-Cost Evaluation of Strip Mining in Appalachia*, Appalachian Resources Project (Knoxville, Tennessee: The University of Tennessee, 1973).

8. Robert Curry, "Reclamation Considerations for the Arid Lands of the Western United States," statement prepared for United States Congress, House Committee on Interior

and Insular Affairs, *Regulation of Surface Mining Operations*, Part 2 (Washington, D.C.: Government Printing Office, 1973), pp. 1006-15.

9. Thadis Box et al., *Rehabilitation Potential of Western Coal Lands*, National Academy of Sciences (Cambridge: Ballinger Publishing Company, May 1974).

10. Energy Policy Project of the Ford Foundation, as reported in *Exploring Energy Choices* (Washington, D.C.: Ford Foundation, 1974), p. 25.

11. Utah's production is almost entirely mined underground.

12. "Western Coal Development," *Coal Age*, May 1974, p. 90. By contrast, average rainfall in Appalachia can be up to 45 inches and in other Western areas, 14-16 inches, as reported in Council on Environmental Quality, *Coal Surface Mining and Reclamation* (Washington, D.C.: Government Printing Office, 1973), p. 14.

13. An excellent summary of available surface-mining and reclamation techniques is given in Council on Environmental Quality, *Coal Surface Mining and Reclamation* (Washington, D.C.: Government Printing Office, 1973). See Chapter 1, pp. 11-35. The figures here are from p. 28.

14. As reported in Charles River Associates, *The Economic Impact of Public Policy on the Appalachian Coal Industry and the Regional Economy* (Cambridge: Charles River Associates, 1973), p. 96.

15. As reported in Richard L. Gordon, "Environmental Factors in Energy Production and Use," "Energy Supply Project" (unpublished manuscript on file, Resources for the Future, Washington, D.C., 1973), p. 42. Whether these figures are for current proposed levels or earlier state regulations is unclear.

16. Charles River Associates, op. cit., no. 2, pp. 3-15.

17. Council of Environmental Quality, op. cit., pp. 35-48, for a discussion of state statutes.

18. Congressional Research Service, *Factors Affecting the Use of Coal in Present and Future Energy Markets* (Washington, D.C.: Government Printing Office, 1973), p. 25.

19. This discussion is based on Everett F. Goldberg et al., *Legal Problems of Coal Reclamation*, Environmental Protection Agency (Washington, D.C.: Government Printing Office, 1972), pp. 63-82.

20. Ibid., p. 82.

21. Council on Environmental Quality, op. cit.

22. Box et al., op. cit.

23. As reported in Massachusetts Institute of Technology Energy Policy Group, "Energy Self-Sufficiency: An Economic Evaluation," *Technology Review*, May 1974, p. 38.

24. Ibid.

25. From *Newsweek*, August 5, 1974, cited in *Environment* (Washington, D.C.: Environmental Policy Center, August, 1974).

26. Ibid. Note that this is a somewhat strong assertion given that the switch from natural gas to coal in the Eastern states due to supply problems was the major original impetus to surface production in the West.

27. Based on laws as discussed in Council on Environmental Quality, op. cit., pp. 35-48. Montana at present has few stringent regulations.

28. See particularly Schmidt-Bleek et al., op. cit., pp. 13-20.

29. Robert Bohm et al., *Benefits and Costs of Surface Coal Mine Reclamation in Appalachia*, Appalachian Resources Project (Knoxville, Tennessee: The University of Tennessee, 1974), p. 6.

30. Ibid., p. 5.

31. As presented by David Hackett, engineer working on reclamation costs for the Appalachian Resources Project. His main effort at present is evaluating these current Pennsylvania projects.

32. *Business World*, August 3, 1974, p. 46.

33. As suggested by Schmidt-Bleek et al., op. cit., pp. 20-21.

34. Hubert Hagen, statement prepared for United States Congress, Senate, Committee on Interior and Insular Affairs, 93rd Congress, 1st Session, Hearings. In *Regulation of Surface Mining Operations*, Part 1 (Washington, D.C.: Government Printing Office, 1973), pp. 764-93.

35. See *Coal Age*, May 1974, p. 91.

36. *Coal Age,* op. cit., p. 9.

37. Costs which do not vary with deliveries have no effect on the minimum cost solution or shipment patterns in these linear programming models. In particular, costs which are uniform across regions have this effect. See James Henderson, "A Short Run Model of the Coal Industry," *Review of Economics and Statistics* 37 (1955): 337.

38. United States Department of Interior, *Surface Mining and Our Environment* (Washington, D.C.: Government Printing Office, 1967), pp. 54, 83.

39. Council on Environmental Quality, op. cit., particularly pp. 49-82.

40. For the summary of their key assumptions for each case, see ibid., p. 75.

41. Council on Environmental Quality, op. cit., p. 55.

42. Ibid., p. 66.

43. Cases where environmental efforts to control a particular pollution problem have resulted in stimulating an alternative problem are discussed in George Hagevik, *Decision Making in Air Pollution Control* (New York: Praeger, 1970).

44. Bohm et al., op. cit., p. 1.

45. This is the conclusion also reached by Council on Environmental Quality, op. cit., pp. 59-60.

46. In 1972, output per man per day averaged almost 36 tons at strip mines, only 12 tons at underground mines.

47. Council on Environmental Quality, op. cit., p. 67. We note that it is often ambiguous as to what truly is the base for calculating "total" resources.

48. Hagen, op. cit., p. 790.

7

THE SULFUR
EMISSIONS TAX

We have noted that a sulfur emissions tax has been proposed as an alternative to a system of emissions standards. There have been two main pieces of legislation proposed. The Pure Air Tax Act was proposed during President Nixon's administration.[1] It called for a tax of $300 per ton on sulfur emissions for those areas that do not meet primary air-quality standards and a tax of $200 per ton for areas that have met the primary standards but not the secondary standards. There would be no tax where both primary and secondary standards are met.

As noted in a Congressional Research Service study,[2] current proposals for such a tax favor a uniform national rate such as suggested by Senator William Proxmire.[3] A differential tax rate may give industrial polluters incentives to relocate their plants to regions with clean air quality and a lower tax rate. Given the importance of industrial tax revenues to local governments, such switching could be undesirable.[4] Furthermore, in order to insure or even improve the effectiveness of the tax as industry and the economy expand, it has been suggested that the tax rate per pound of emissions should increase over time.[5] In the Proxmire proposal, the sulfur emissions tax would increase at a rate of 5 cents per pound of emitted sulfur over a four-year period. Specifically, the rate would start at 5 cents per pound of emitted sulfur and increase to a final tax level of 20 cents.

It should be noted that all major FPC power regions that at present rely on coal-fired generation would be taxed at the higher rate under the proposed differential tax system.[6] The four largest of these power regions—Northeast Central, South Atlantic, Southeast Central and Middle Atlantic—are responsible for 86 percent of coal used in electric generation.[7] The West, with its traditional reliance on natural gas as the major generator fuel and the availability of lower sulfur coals for electric generation, would probably not be placed in the higher bracket. Even in regions that rely heavily on natural gas for steam electric generation, such as the Southwest Central area, the higher rate would apply as a result of industrial activity. Therefore, all states either relying at present on coal-fired generation or planning significant additions to coal capacity probably would be taxed at the highest rate. Even in the West, where the states have generally moved toward rather strict emissions controls, any significant switch in generation from natural gas to coal could result in higher taxes. This is particularly important for any tendency to cluster generating plants around a major coal deposit, as we shall see in the next chapter.

At present, most states plan to reduce sulfur emissions in conventional steam electric generation in the near future through fuel choice. In the long run nuclear power and a refined control technology are alternatives, but the effects of the sulfur emissions tax in promoting substitution of lower-sulfur fuels would still be important. We have seen that coal contains more sulfur than other fossil fuels. Therefore, we will now investigate the possible substitution effects a tax may have on coals of different sulfur content and the subsequent regional effects on the coal industry.

In particular, we modify the objective function (3.4) to:

$$\text{Minimize} \sum_i \sum_j \sum_k \sum_l (C_{ij} + t_{il})X_{ijkl} + \phi(b_{ijk}S_{ijk})X_{ijkl}$$

where ϕ is the level of the sulfur emissions tax. The tax alters the cost to the user of a ton of coal from any region by the emitted pounds of sulfur per million Btus in that ton times the appropriate tax. The term $b_{ijk}S_{ijk}$ represents the sulfur emissions in the model that is to be taxed.

We have seen over recent years that the average sulfur content of coal, as delivered to utilities, has been 2.9 percent. In particular, deliveries from major centers of coal production for utility consumption in the Midwest—Illinois and Western Kentucky—have averaged 3.4 percent sulfur content. We wished to investigate the effects of the sulfur emissions tax on the use of high-sulfur coal. We will define high-sulfur coal as that containing 3 pounds of sulfur or more per million Btu. Consider a coal with 3.5 pounds of sulfur per million Btu and 11,000 Btu per pound, the characteristics typical of some Illinois production. A ton of such coal would generate almost 22 million Btu of energy value, but with sulfur emissions of 77 pounds. At a tax rate of 10 cents, this would add $7.70 per ton to the cost of this coal use. Thus the sulfur emissions tax can contribute to significant cost differentials for the user among different coal deposits.

There have been four different levels proposed for such a tax, and we will consider the effects of each of these. Table 7.1 presents the regional shipments of high-sulfur coal (above 3 pounds) at the various tax levels. Table 7.2 compares the use of these coals to the solution with no sulfur emissions tax. We see that the sulfur emissions tax can significantly affect the utilization of high-sulfur coal. At a tax rate of 15 cents per pound of emitted sulfur the use of high-sulfur coal is reduced by 72 percent.

TABLE 7.1

Regional Shipments of High-Sulfur Coal with a Sulfur Emissions Tax (millions of tons)

Region	Sulfur Emissions Tax (dollars per pound)				
	0.00	0.05	0.10	0.15	0.20
Northern Appalachia	27.020	25.640	16.500	1.140	1.140
Central Appalachia	0.510	0.000	0.000	0.000	0.000
Southern Appalachia	0.047	0.047	0.047	0.047	0.047
Appalachia, total	27.120	25.690	16.550	1.190	1.190
Midwest	69.940	57.780	35.010	25.820	24.400
West*	—	—	—	—	—
United States	97.060	73.470	41.560	27.010	25.590

*Western producers ship no coal with 3 pounds or more of sulfur per million Btu.
Source: Compiled by author.

TABLE 7.2

Percentage of High-Sulfur Coal Shipped in Relation to a No-Emissions-Tax System

Region	Sulfur Emissions Tax (dollars per pound)			
	0.05	0.10	0.15	0.20
Appalachia	95	61	4	4
Midwest	83	50	37	35
West*	—	—	—	—
United States	77	44	28	26

*No high-sulfur coal shipped.
Source: Compiled by author.

As we might expect, much of the regional impact of such a sulfur emissions tax depends upon the level at which Western surface production can overcome its relative transport-cost disadvantage as the tax increases. Table 7.3 shows the regional extraction of coal at the different tax levels. The regional interaction between Midwestern and Western production is particularly strong. At a 10-cent tax rate, Midwest surface production falls by 30.7 million tons, while Western

TABLE 7.3

The Regional Effects of a Sulfur Emissions Tax on Coal Extraction (millions of tons)

	Sulfur Emissions Tax (dollars per pound)				
Region	0.00	0.05	0.10	0.15	0.20
Northern Appalachia	165.7	162.5	155.6	140.2	140.2
Central Appalachia	122.1	129.6	135.7	135.7	135.7
Southern Appalachia	20.1	20.1	20.1	20.1	20.1
Appalachia, total	307.9	312.2	311.4	296.0	296.0
Midwest	154.0	142.9	123.8	114.6	113.5
West	41.9	48.9	71.8	102.9	105.2
United States	503.8	504.0	507.0	513.5	514.7

Source: Compiled by author.

TABLE 7.4

Effects of Additional Western Production with a Sulfur Emissions Tax at 20 Cents per Pound (millions of tons)

Region	Shipments Without Additional Capacity	Shipments With Additional Capacity	Change in Shipments	Level of High-Sulfur Shipments
Midwest	113.5	92.6	−20.9	11.1
West	105.2	129.0	+23.8	—
United States	514.7	517.6	+2.9	12.3

Source: Compiled by author.

shipments increase by 28.7 million tons. The largest absolute change in the percentage utilization of high-sulfur coal occurs as we move from a 5-cent to the 10-cent tax rate, a decrease of 43 percent. This is so particularly for Midwestern producers.

As we move to a 15-cent tax rate, however, shipments of high-sulfur coal from Northern Appalachia fall nearly to zero. As seen in Table 7.2, the decline in Northern Appalachia from a 15-cent tax decreases the overall use of Appalachian high-sulfur coal to only 4 percent relative to shipments with no sulfur emissions tax. The higher-sulfur coals in Northern Appalachia have an average of almost 13,000 Btu per pound as delivered to utilities, and one ton of coal with 3.5 pounds of emitted sulfur per million Btu would contain almost 91 pounds of sulfur, even though it had extremely high energy value. At a tax rate of 15 cents, the sulfur emissions penalty in utilization of this coal would be $13.65 per ton. At this rate, the relative delivered cost advantage of this coal in FPC regions of New England and the Middle Atlantic is lost, compared to higher delivered cost and lower-sulfur underground coal from Central Appalachia. Since Eastern utilities can outbid Midwestern markets for lower-sulfur Appalachian coal, the Midwest's reliance on Western coal increases.

Compared to their use when no emissions tax is in effect, Western shipments increase greatly at a tax rate of 20 cents, by almost 61 million tons or 250 percent! This is what we would expect, for in the example we considered earlier, the increase in user cost of higher-sulfur Midwestern coal would be $15.40 at this tax rate. Western producers can easily overcome their delivered cost disadvantage in Midwestern markets. Indeed, the model would use even more Western coal than indicated in Table 7.3, had we not reached the short-run capacity limits specified for Western coal in the basic model at the 20-cent tax rate.

As an exercise in pure simulation with the model, we considered what would occur if there were no short-run limits on production potential from Western fields. These effects are shown in Table 7.4. The deliveries of Western coal into the Midwestern markets would increase by almost 24 million tons, and as a result, local Midwest production would fall by an additional 21 million tons. The overall use of high-sulfur coal would be only slightly higher than 8 percent of capacity. We can conclude that such a tax would give Western producers a large competitive advantage over local production in Midwestern markets, even though unlimited capacity is not likely to occur. Of course, if production from lower-sulfur deposits in Midwestern and Appalachian fields were also unlimited, Western production would once again face a delivered cost disadvantage and a tax would not increase its production at all. One has to be careful, however, in drawing conclusions from purely simulated results.

The benefits derived by society from a sulfur emissions tax would be the decrease in emissions and the damage they cause. Table 7.5 gives the levels of emitted sulfur at each tax rate. Significant decreases do occur. At a 15-cent tax, for example, emissions are reduced by 19 percent compared to the no-tax situation. If damages from sulfur emissions are approximately 30 cents per pound as proposed earlier, the decreases in overall damages are $877.5 million at the 10-cent tax level and $1.264 billion at the 20-cent tax rate.

Such a sulfur emissions policy affects costs in two areas. The tax, designed to improve air quality, can have significant land-use impacts. The interaction between the sulfur emissions tax and Western surface mining is particularly significant. The controversy over land-use impacts in Western surface mining at current levels of activity is severe, and a regional shift of a large percentage of

TABLE 7.5

Total Effects of the Sulfur Emissions Tax

Effect	Sulfur Emissions Tax, dollars per pound				
	0.00	0.05	0.10	0.15	0.20
Total emissions million tons	11.41	10.72	9.95	9.33	9.30
Benefit of abatement, millions of dollars	—	415.0	877.0	1,245.2	1,264.2
Total tax revenue, billions of dollars	—	1.072	1.989	2.800	3.721

Source: Compiled by author.

TABLE 7.6

Quasi-Rents for Selected Producing Areas with a 20-Cent-per-Pound Sulfur Emissions Tax

Region	Quasi-Rents, dollars per ton, at pounds Sulfur per million Btu				
	0-0.6	0.6-1.0	1.0-2.0	2.0-3.0	3.0 or greater
Eastern Kentucky					
Surface	14.96	11.81	0.86	0.35	0
Underground	14.32	11.46	0.83	0.32	0
Northern West Virginia					
Surface	12.63	13.77	9.26	4.84	0
Underground	16.56	12.99	9.15	3.32	0
Wyoming					
Surface	7.52	6.42	—*	—	—
Underground	—	2.96	—	—	—
Colorado					
Surface	10.89	9.07	6.48	—	—
Underground	8.84	4.38	—	—	—

*No such coal shipped.
Source: Compiled by author.

Midwestern production to the West would only add to the problem. The tax also provides a stimulus to surface-mine investment in Appalachia in low-sulfur production. As seen in Table 7.6, the low-sulfur deposits in Appalachia earn the greatest quasi rents under a sulfur emissions tax. A sulfur emissions tax, therefore, would stimulate production and investment potential in Western and steep Appalachian deposits where the effect of surface mining on land use is most controversial for the coal industry at present.

The second area affected by the emissions tax policy is the delivered cost of coal to utilities. Table 7.7 presents the levels of these costs at the different sulfur emissions tax rates. The largest changes are in the Midwest. The average increase in the Midwestern market is 48 cents per million Btu at the 15-cent rate and 66 cents at the 20-cent rate. Given the concentration of high-sulfur coal from Midwestern producers, this is what we would expect. In the East, however, the cost increases are only slightly less than in the Midwest, with an average change at the 15-cent and 20-cent rates of 46 and 63 cents per million Btu, respectively. Western utilities, relying on local coal with relatively low-sulfur levels and production costs, generally show no cost effects from either reclamation or sulfur emissions regulations. The impact of such policies falls mainly on the Midwest and East. With a sulfur emissions tax, however, Western utilities do experience cost increases in coal-fired generation at an average for the whole region of 23 cents per million Btu at a 15-cent tax rate and of 40 cents per million Btu at the 20-cent tax. The absolute levels of Western utility coal costs are significantly less than in other areas of the country and this relative differential, as seen in Table 7.7, increases with the higher tax levels.

We have noted that one purpose of a sulfur emissions tax would be the incentive for long-run development of better emission control technology by utilities. The total taxes collected under the alternative sulfur emission tax rates are given in Table 7.5. Even though total sulfur emissions decline as tax levels rise, overall increases in tax revenues still result. Between the 5-cent and 20-cent tax rate, revenues increase by a factor of 3.5. If tax revenues, less administrative costs, were allocated for additional research to lower the sulfur levels of coal either through synthetic fuels, new coal cleaning technology, or stack-gas scrubbing techniques, the research fund would be considerable. This research program could be administered directly through the federal government, assuming all taxes from the utilities were collected by a federal agency, or perhaps through utilities themselves, which in turn might be allowed to deduct the costs of research up to the total level of their tax.

Our general conclusion is that an attempt to improve air quality through a sulfur emissions tax could cause significant environmental degradation in land use. As we found with other policies we have investigated, regional interaction and overall land-use impacts must be considered in evaluating any policy, for these effects can be significant.

TABLE 7.7

Changes in the Delivered Cost of Stream Coal
Due to a Sulfur Emissions Tax

State	Sulfur Emission Tax, dollars per pound, at dollars per million Btu				
	0.00	0.05	0.10	0.15	0.20
Alabama	0.49	0.65	0.82	0.93	1.10
Arizona	0.36	0.40	0.46	0.56	0.73
Arkansas	–	–	–	–	–
California	–	–	–	–	–
Colorado	0.40	0.43	0.56	0.70	0.87
Connecticut	0.60	0.73	0.89	1.06	1.23
Delaware	0.50	0.62	0.78	0.95	1.13
District of Columbia	0.52	0.64	0.80	0.97	1.15
Florida	0.60	0.72	0.89	1.05	1.22
Georgia	0.49	0.61	0.78	0.95	1.12
Idaho	–	–	–	–	–
Illinois	0.45	0.61	0.76	0.93	1.11
Indiana	0.40	0.57	0.74	0.92	1.09
Iowa	0.50	0.57	0.69	0.83	1.01
Kansas	0.50	0.57	0.69	0.80	0.98
Kentucky	0.39	0.55	0.73	0.90	1.08
Louisiana	–	–	–	–	–
Maine	–	–	–	–	–
Maryland	0.52	0.64	0.80	0.97	1.15
Massachusetts	0.63	0.75	0.92	1.09	1.26
Michigan	0.54	0.66	0.83	1.00	1.17
Minnesota	0.59	0.63	0.71	0.81	0.98
Mississippi	0.54	0.66	0.76	0.86	1.04
Missouri	0.42	0.58	0.73	0.90	1.08
Montana	0.31	0.35	0.43	0.54	0.71
Nebraska	0.49	0.52	0.65	0.79	0.96
Nevada	0.52	0.56	0.64	0.79	0.92
New Hampshire	0.64	0.77	0.93	0.74	0.92
New Jersey	0.57	0.69	0.85	1.07	1.20
New Mexico	0.30	0.34	0.38	0.48	0.66
New York	0.58	0.70	0.87	1.03	1.21
North Carolina	0.49	0.61	0.78	0.95	1.12
North Dakota	0.40	0.47	0.55	0.62	0.79
Ohio	0.48	0.62	0.79	0.95	1.13
Oklahoma	0.49	0.56	0.65	0.75	0.92
Oregon	–	–	–	–	–
Pennsylvania	0.44	0.56	0.73	0.90	1.07
Rhode Island	0.63	0.75	0.92	1.09	1.26
South Carolina	0.53	0.65	0.82	0.99	1.16
South Dakota	0.53	0.60	0.68	0.76	0.92
Tennessee	0.39	0.51	0.68	0.85	1.02
Texas	0.54	0.58	0.67	0.77	0.95
Utah	0.39	0.43	0.51	0.62	0.79
Vermont	0.63	0.75	0.92	1.09	1.26
Virginia	0.52	0.65	0.81	0.98	1.15
Washington	–	–	–	–	–
West Virginia	0.41	0.54	0.70	0.87	1.05
Wisconsin	0.52	0.68	0.81	0.92	1.08
Wyoming	0.31	0.34	0.46	0.60	0.78

Source: Compiled by the author.

The main effects of a sulfur emissions tax are to reduce the use of higher-sulfur coal. It has been suggested that one way to reduce the use of such coal as a generating fuel is simply to ban its use. This has occurred in some urban situations. We have seen, for example, that Chicago uses some Montana coal instead of relying solely on Illinois production like most of the state. But what would be the effect of such a policy as an overall regulation? Reports emphasizing the lack of lower-sulfur coals and the granting of variances for utilities to burn high-sulfur coal even in urban areas would lead us to believe that such a policy on a short-term basis would be unfeasible.[8]

In our model, a ban on the use of high-sulfur coal simply cannot be upheld, since there is not enough low-sulfur coal available to compensate for the production lost. Over 97 million tons of high-sulfur coal, produced mostly in the Midwest, cannot be replaced under present capacities by substitution of low-sulfur coals. The basic solution in our model shows 19.3 percent of total coal used is high-sulfur coal. One might ask if this figure is perhaps too high; that actual use is so much less than in the basic model (97.2 million tons), that we overestimate the importance of high sulfur output. If we aggregate the data available from the FPC,[9] which list all coal deliveries to utilities by sulfur content, we find that over the last half of 1972 and through 1973 approximately 26 percent of total steam electric coal was high sulfur. Alternate outputs to compensate for an overall ban of such production for coal-fired utility boilers would be impossible.

Table 7.8 shows the simulated increases in local production required to replace high-sulfur coal in the basic solution without changing the relative regional production levels. The greatest expansion would be required in the Midwest, where any deficiency in supplying this additional tonnage by Midwest

TABLE 7.8

Regional Increases in Coal Production to Compensate for High-Sulfur Coal Losses

Region	Percent Increase in Other Capacity	Required Additional Tonnage, millions of tons
Appalachia	9.68	27.1
Midwest	83.21	69.9
West*	—	—
United States	23.82	97.0

*No increase required.
Source: Compiled by author.

producers would require the use of Western coal. As we have seen, this would add approximately 22 cents per million Btu to the average delivered costs of coal in the Midwest utility market. Overall, net sulfur emission levels would decrease by 12 percent, almost 2.135 million tons, if the use by the utilities of all high-sulfur coal was eliminated and other coal substituted. This is a reduction comparable to that under the sulfur emissions tax at its highest levels. However, new sources for 97 million tons of coal would have to be found, a difficult task considering the current time lag for expanding existing underground and surface mining or establishing new mines.

The use of taxes for limiting environmental pollutants, an administrative alternative to the enforcement of an emissions standard system, functions in the same way as in limiting discharges of air or water effluents. Let us consider this issue in terms of our linear programming model of production. We have utilities in our regional system attempting to minimize their coal-fired generating costs while meeting required electrical output and yet we have constrained them to a sulfur emissions standard. In matrix notation, our problem is to:

$$\text{Minimize } CX \tag{7.1}$$

$$\text{such that } AX \leqslant \bar{e}$$
$$DX \geqslant \bar{q}$$

where A is our matrix of sulfur emissions with $\bar{e}$ the emissions standard, and D is the matrix of Btu values with $\bar{q}$ the required generating levels.

The solution to (7.1) could provide the same solution to the following problem, where we shall call t the emissions tax, if two conditions were satisfied.

$$\text{Minimize } CX + tAX \tag{7.2}$$

$$\text{such that } DX \geqslant \bar{q}$$

First, the optimal solution to (7.2) must be a feasible solution to (7.1), that is, the emissions tax t meets the environmental input constraint. Secondly, the value of the objective function for the optimal solution to (7.2) less tax collections must be equal to the value of the objective function in the optimal solution to (7.1), that is, the calculated tax causes production to meet the environmental resource limit at minimum costs. It has been shown that the shadow prices on the environmental constraints in (7.1) will allow us to calculate the required value of the tax that satisfies the two conditions simultaneously.[10] More generally, the objective function and constraint system need only to be concave functions over the nonnegative orthant for this result to occur. The linear system considered here is a special case of this result.[11] The implication for public policy is that in most cases there is a theoretical equivalence between a system of emissions standards and a corresponding set of emissions taxes.

TABLE 7.9

Emission Standards Corresponding to the 20-Cent-per-Pound Sulfur Emissions Tax and Those with a 1.65-Pound Sulfur Emission Standard

State	Sulfur Emissions Standard, pounds per million Btu	
	With 20-Cent-per-Pound Sulfur Emissions Tax	With Maximum Allowable Level of 1.65 Pounds
Alabama	1.19	1.65
Arizona	0.80	0.80
Arkansas*	—*	—
California	—	—
Colorado	0.46	0.67
Connecticut	1.22	1.65
Delaware	1.50	1.65
District of Columbia	1.50	1.65
Florida	0.53	1.65
Georgia	1.61	1.65
Idaho	—	—
Illinois	2.20	1.65
Indiana	2.81	1.65
Iowa	0.36	1.65
Kansas	0.80	1.65
Kentucky	1.95	1.65
Louisiana	—	—
Maine	—	—
Maryland	2.31	1.65
Massachusetts	0.57	1.65
Michigan	0.72	1.65
Minnesota	0.97	1.65
Mississippi	0.80	1.65
Missouri	2.81	1.65
Montana	0.80	1.65
Nebraska	0.30	1.65
Nevada	0.31	0.31
New Hampshire	0.30	1.65
New Jersey	1.50	1.65
New Mexico	0.80	0.80
New York	1.56	1.65
North Carolina	0.80	1.65
Ohio	2.29	1.65
Oklahoma	—	—
Oregon	—	—
Pennsylvania	1.80	1.65
Rhode Island	—	—
South Carolina	1.50	1.65
South Dakota	1.50	1.65
Tennessee	0.99	1.65
Texas	—	—
Utah	0.30	0.30
Vermont	0.30	1.65
Virginia	1.36	1.65
Washington	—	—
West Virginia	1.27	1.65
Wisconsin	1.25	1.65
Wyoming	0.30	0.80

*No coal used in electric generation.
Source: Compiled by the author.

A sulfur emissions standard of 1.65 pounds of sulfur per million Btu would result in an overall emission level of 9.30 million tons of sulfur. This would represent a reduction in total sulfur emissions compared to no standard of 2.11 million tons, or almost 4.22 million tons of sulfur oxide pollutants reduced. This decrease of approximately 19 percent is exactly the reduction caused by a sulfur emissions tax at a rate of 20 cents per pound. This is the lowest overall emissions standard which is feasible in the short run, according to this study.

Table 7.9 shows the sulfur emissions standards which would give the same regional production and shipment pattern as under the 20-cent sulfur emissions tax. It also gives the resulting average sulfur emissions which occur under the 1.65 standard. Even though the total level of sulfur emissions under the two policies is the same, the regional implications are quite different. The Western states are relatively unaffected by the tax since they are able to satisfy their requirements with local low-sulfur production which brings their average emission level below the 1.65-pound standard. The system of standards allows a uniform reduction in average emissions across all regions, while the sulfur emissions tax creates a differential standards system with wide variability. Indeed, in the large steam coal producing states like Ohio, Illinois, Indiana, etc., the average level of sulfur emissions still exceeds 2.00 pounds per million Btu under a 20-cent tax. Therefore, a policy which is concerned with the average level of emissions from power plants rather than total emissions would be achieved better by a uniform source standard rather than a tax.

NOTES

1. Council on Environmental Quality, *The President's 1972 Environmental Program* (Washington, D.C.: Government Printing Office, 1972).

2. Congressional Research Service, *Factors Affecting the Use of Coal in Present and Future Energy Markets* (Washington, D.C.: Government Printing Office, 1973), pp. 34-35.

3. See United States Congress, Senate, statement by Senator William Proxmire, 92nd Congress, 2nd Session, *Congressional Record* 118 (5): 276-79 (Washington, D.C.: Government Printing Office).

4. These institutional considerations in public policy are considered in Robert Bish, *The Public Economy of Metropolitan Areas,* Markham Public Finance Series (Chicago: Markham Publishing Co., 1971).

5. Congressional Research Service, op. cit., p. 35.

6. Based on FPC Form 423 data.

7. Derived from Federal Power Commission, *Monthly Report of Cost and Quality of Fuels for Steam Electric Plants* (Washington, D.C.: Government Printing Office, December 1973), Table 4.

8. See Executive Office of the President, Office of Emergency Preparedness, *The Potential for Energy Conservation: Substitution for Scarce Fuels* (Washington, D.C.: Government Printing Office, 1973), or see statement by Russell Train, administrator of EPA, "Coal Burning Power and the E.P.A.," Washington *Post*, September 5, 1974.

9. Specifically, Federal Power Commission, Form 423 data which require utility reporting of cost and quality of all fossil fuels on a monthly basis.

10. Particularly in A. Kneese and B. Bower, *Managing Water Quality: Economics, Technology, and Institutions* (Baltimore: Johns Hopkins Press, 1968), Chapter 7, and W. Baumol and W. Oates, "The Use of Standards and Pricing for the Protection of the Environment," *Swedish Journal of Economics* 73 (1971): 42-54.

11. An empirical estimation of taxes corresponding to standards of nitrogen fertilizer use in a linear programming model of the agricultural sector can be found in Lawrence Abrams and James Barr, "Corrective Taxes for Pollution Control," *Journal of Environmental Economics and Management*, 1, 1974, pp. 296-318.

CHAPTER

8

WESTERN COAL

Within the West, there are controversies which have severely affected the potential for coal-fired generation. Adequate reclamation of acreage that has been disturbed by surface mines supplying utility coal has been the most publicized controversy. Even if we assume that complete reclamation is possible, problems remain. We have seen that the water requirements for the disposal of waste heat from steam electric plants can significantly affect decisions concerning the location of power plants. In relatively arid states, water needed to supply evaporative cooling towers, etc., may place an undesirable drain on available surface- and ground-water sources. The National Academy of Sciences has concluded that adequate water resources are not available for large-scale development of Western coal for gasifying and liquefying industries or generating electric power.[1]

In order to provide water supplies to energy conversion facilities in the West, some natural water courses would be diverted and aqueducts constructed. A Bureau of Reclamation study in the Montana-Wyoming region has considered the diversion of 2.6 million acre feet of water resources per year in that region.[2] The study concludes that such a policy would involve a significant proportion of natural river flow in the area and could substantially affect the region's watershed.

Most power plants are planned for an operating life of less than 50 years. As Box has stated, "What will be done with the proposed aqueducts, dams, and water flow when the coal economy no longer needs them? Does society wish to commit water resources to an ephemeral industry at a particular place?"[3] Reaction from local and state political powers in the West to the prospect of uncontrolled surface mining indicates that tampering with water resources would hardly be encouraged.

It might be expected that any large-scale development of coal-fired generation in the West would be concentrated in the vicinity of the major coal-producing deposits. Long-distance transmission lines for several plants and the pooling of power could reduce the individual plant-distribution costs. Rail transport is less extensively developed than in the East, and the rugged topography between major rail centers necessitates long, winding routes. In the West, coal deposits can lie in remote, sparsely inhabited areas and a necessarily extensive rail system connecting them with relatively distant urban centers has not been developed.

The greatest development of Western power plants at the source of coal deposits has been the Four Corners Project, located at the boundaries of Nevada, Arizona, New Mexico, and Colorado, where large coal reserves made the project feasible. This concentration of power plants has been strongly attacked.[4] Four Corners, representing the largest technologically feasible coal-fired capacity, has reduced air quality in the area with particulate and sulfur emissions. The utilities involved were able to avoid the rigorous California environmental standards by locating the power plants outside their service area. The U.S. Supreme Court has since ruled that one region's environmental difficulties cannot be solved at the expense of reducing the environmental quality of another region. This has limited other large mine-mouth projects, like the Four Corners venture, planned by several utilities. In 1973 the Department of Interior refused to allow construction of the final, and largest, of the planned additions, even though the Davis Dam on the Colorado River was capable of meeting critical water requirements in this area.[5]

The West Coast utilities, owners of most mine-mouth projects, are reported to be increasingly pessimistic about the entire mine-mouth concept.[6] After reviewing developments in Wyoming as well as Four Corners, Gordon concluded that, "In sum, it would appear that a relatively rare combination of circumstances is required to encourage construction of plants outside a company's service area. All present and planned examples involve supply of areas in which land is expensive, environmental pressures are severe, extensive transmission facilities exist and fuel transport is fairly expensive."[7]

The availability of Western coal for surface mining can be seen in Table 8.1. Reserves of this strippable coal are generally recoverable at a rate of 80 percent.[8] Of approximately 57 billion tons of strippable reserves in the West, about 160 million tons had been surface mined through 1973. Attempts have been made to estimate how much of these reserves are actually mineable. By using assumptions concerning allowable ratios of overburden to seam thickness, allowable mining conditions and quality of the coal, etc., one could alter the estimates in any direction. However, the Bureau of Mines' estimate of these Western strippable reserves is often cited as the most consistent and careful.[9] This study indicates that 25.7 billion of the West's strippable reserves can be recovered satisfactorily at low cost.

TABLE 8.1

Western Reserves by State of Strippable Coal Resources with Range of Overburden Less Than 100 Feet

State*	Surface-Mining Coal, Original Resources, millions of tons
North Dakota	15,000
South Dakota	400
Montana	23,000
Wyoming	13,000
Idaho	0
Total, Northern Rocky Mountains	51,400
Colorado	1,200
Utah	300
Arizona	400
New Mexico	3,000
Total, Southern Rocky Mountains	4,900
Washington	500
Oregon	—
California	0
Total, West Coast	500
Total, Western	56,800

*With regional groupings corresponding to those in Table 4.3.

Source: Paul Averitt, *Stripping Coal Resources of the United States—January 1, 1970*, U.S. Geological Survey, Bulletin 1322, Washington, D.C., 1970, p. 23.

One issue concerns the extent to which Western coal costs would be affected as deposits are depleted and increasingly higher overburden ratios drive up production costs. Bureau of Mines' data indicate that this situation is not expected to arise in the near future, since reserves of 13.6 billion tons are estimated in the Wyoming-Montana Powder River Basin at current overburden ratios. Assuming a typical mine life of 20 years, the Powder River Basin could produce 680 million tons per year before depletion would necessitate the mining of new deposits with higher overburden ratios. With a time horizon of 30 years, total Western production could reach almost 850 million tons per year. In this 20-to-30-year period, however, it might be reasonable to expect that advances in productivity would offset higher costs due to greater overburden ratios.

Any large development of Western reserves in the future would require migration of manpower and extensive mine and equipment investment. Thus, in Table 8.2 we give some indication of the life of reserves at more reasonable growth rates. Even if we consider the lower figure, 72 years represents a considerable life span for electric-utility resources. Gordon has calculated that, if all of these reserves were used entirely for the production of synthetic gas, they could produce 9.5 trillion cubic feet of gas per year for 30 years.[10] By comparison, annual output of natural gas has been over 20 trillion cubic feet during the past few years. We will note, therefore, that although these coal reserves are not large enough to be our major energy resource over the long run, they are a significant one.

On the basis of percent by weight, Western coal is generally classified as low sulfur. There is, however, some uncertainty about the energy value of these Western coals and, hence, about their sulfur content as measured per million Btu. For instance, one ton of coal, of 1.5 percent sulfur by weight, and containing 30 pounds of sulfur, would have a greater level of sulfur emissions if it produced only 8,000 Btu per pound than if it generated 12,000 Btu per pound. As the MIT Energy Policy Study Group concluded "an additional uncertainty with respect to Western reserves is whether the low sulfur supplies from this region satisfy environmental standards."[11] The FPC data on Western coal shipments indicate that over 71 percent of coal shipped contains less than 1 pound of sulfur per million Btu, and over 85 percent less than 1.5 pounds of sulfur per million Btu.[12] These data indicate that current Western deposits, and those reserved for future use, are indeed of sufficient energy value to allow us to consider them as truly low-sulfur deposits. Generally, the average energy value of Western surface shipments is over 9,000 Btu per pound.

Federal leasing policies can be crucial for Western coal development since 35-40 percent of the deposits are located on federal land.[13] Furthermore, federally owned deposits often lie next to private deposits in a "checkerboard" pattern which could limit major mining development and large tract acquisition

TABLE 8.2

Approximate Life of Reserves at Percent Growth Rate, years

	1970 Production[a]	3 Percent	5 Percent
Total resources	2,272	223	156
Mineable resources[b]	1,049	103	72

[a]0 percent.
[b]Estimate by Bureau of Mines.
Source: Averitt, op. cit., p. 23.

unless federal leases were granted. Under the Mineral Leasing Act of 1920, coal mining and acquisition rights are allowed on these public lands on a lease basis with regulated minimum annual rentals and royalties.[14] This leasing system, designed to encourage mineral exploration and development on public lands, has been criticized for two reasons. The Interior Department has been given arbitrary authority in administering the leasing system, and, since there has been no overall plan for the development of these resources, it has allowed extensive leasing that is speculative in nature.

The Interior Department has initiated significant modifications of its coal-leasing policy. A partial moratorium on leases is currently in effect until completion of a regional development plan for the "orderly development of the coal in these states with adequate environmental protection."[15] The proposals for altering the coal-leasing system, however, are not generally seen as limits to Western coal development. A U.S. Senate resolution in 1972 froze all further leasing of rights in Montana until federal surface-mining legislation had been passed.[16]

The effects of public policy on Western coal costs should not seriously affect the competitive position of Western coal in relation to fossil fuels. Table 8.3 shows the relative prices of the fossil fuels used in electric generation in the West. The effects of the increase in imported oil prices since 1971 is particularly apparent. Reclamation, even at the highest Western expenditure estimates, would add but 2-3 cents per million Btu to the cost of Western coal used in electric generation. Because of the current limited supply of natural gas, the decision for utilities is between oil and coal.[17] The effects of reclamation on coal costs are too small to affect seriously the competitive position of Western coal. Even if we added 20 cents per million Btu to Western coal from a high-sulfur emissions tax, the price of oil would still be twice as high as coal. Imported oil prices may fall in the future, but unless the fall is drastic, coal appears to be the cheapest energy generation source for Western utilities.

Another consideration is the level of Western surface mining which public opinion will allow before development is hindered by local and state action. Most surface-mining reserves occur in relatively remote areas, where the alternative value of grazing is relatively low. Yet the controversy over surface-mine reclamation shows that public opinion can be an important variable in the West. It could significantly affect the use of Western coal by Midwestern utilities for meeting sulfur emissions standards. Sulfur emission policies have boosted the competitive position of Western coal in Midwestern markets in relation to local coal production. Whether Western producers could expand production into new reserve areas might well depend on local response. In the next chapter, we will consider more closely Western coal prospects in the Midwest.

The importance of the railroads to Western coal development is often emphasized. In 1972, 74.4 percent of all coal was shipped by rail, with water and truck shipments 13.2 and 12.4 percent, respectively.[18] Bituminous coal

TABLE 8.3

Unit Costs of Fossil Fuels Used in Steam Electric Generation in the West

| Region | Year | Costs in Cents per Million Btu as Burned | | |
		Coal	Oil	Gas
Mountain	1969	20.6	27.3	27.3
	1970	19.8	28.2	29.3
	1971	20.9	40.4	32.4
	1972	22.7	58.2	35.1
	1973	23.1	101.4	38.7
Percent increase				
	1972/71	+8.6	+44.1	+8.3
	1973/72	+1.8	+74.2	+10.3
Pacific				
	1969	—*	34.5	31.2
	1970	—	36.8	32.4
	1971	—	55.4	34.6
	1972	—	73.9	37.5
	1973	38.8	94.2	42.1
Percent increase				
	1972/71	—	+33.4	+8.4
	1973/72	—	+27.5	+12.3

*No coal used.

Source: Data for 1969-72 from National Coal Association, *Steam Electric Plant Factors, 1972* (Washington, D.C.: National Coal Association, 1973), p. 55. The 1973 figures are from Federal Power Commission, *Monthly Report of Cost and Quality of Fuels for Steam Electric Plants,* December 1973 (Washington, D.C.: Government Printing Office, 1974), p. 22.

traffic has been a major source of railroad revenue, averaging around 25 percent of total revenue in the recent past.[19]

The shipment of coal by rail has been facilitated by the development of the unit train. The unit train consists of a number of large-capacity coal cars, which are never uncoupled and often stop only for crew changes and maintenance. They may be loaded while moving and can release their coal without stopping at power plants with special handling facilities. With the elimination of delays and switchings and with the more efficient use of coal cars, the introduction of the unit train in the early 1960s reduced rates as much as 30 percent in some areas.[20] Indeed, the unit train has eliminated the use of coal pipelines in most areas since 1962.[21] They are extensively employed in long-haul shipments and, if there is increased reliance on Western coal, there may be some question over the capacity and availability of unit trains to handle increased long-haul shipments of Western coal.

The concern over "shortages" of unit train cars and general rail capacity for coal is hardly new. As early as 1923, transport limitations in the railroad system were a bottleneck for coal's potential.[22] More recently, there has been extensive discussion of railroad-car shortages for the shipment of all commodities.[23] The problem of wheat distributors in 1970 was probably more severe than any present coal-car shortage.

The overall potential for expansion of the railroad industry, particularly for unit trains, has been studied. The Federal Railroad Administration believes that utilization of existing hopper cars could improve coal-carrying capacity on short notice by up to 10 percent.[24] Since many hopper cars are kept idle at power plants, serving as coal storage bins, further increases may be possible.[25] Hopper-car producers appear to have a large reserve in capacity, since some have production lines that are idle or are using only one shift. The Ford Foundation's project has estimated that an increase in coal hauling by 50 million tons per year could be handled with little or no difficulty by present railroad capacity.[26] Further increases in coal shipments could also be handled, not only because of the excess capacity in the hopper-car industry, but because increases in locomotive production are also feasible.

A shortage to the economist represents a disequilibrium in supply that should eventually be diminished, particularly if there is no special resource constraint, etc. This assumes that there is no institutional characteristic of the economy which would prevent these adjustments. It has been suggested that the somewhat complex regulations of the Interstate Commerce Commission (ICC) concerning per-diem on rental rates to be paid for the use of railroad cars could be a cause for the freight-car shortage.[27] There has been no system of basic and incentive per-diem charges to stimulate the railroads to acquire and maintain larger supplies of freight cars and to utilize existing cars more efficiently.

Over the long run, the new technology for producing synthetic fuels from coal could be important. The technology of oil shale could act as a competitive substitute for Western coal liquefaction. However, both processes are subject to significant water requirements, and, as we have seen, water resources are limited in the West. Although the water requirements for oil-shale recovery are only one-third of those for coal liquefaction, the water resources needed for any large development may be inadequate.[28] The most highly developed technique for oil-shale recovery, that is, pulverizing terrain in which it occurs and then "restacking" the hills, has an aesthetic effect which is difficult to measure. Revegetation of the spoil in less than 10 years may be impossible. The in situ process, which attempts to recover the oil from the shale in place, is not as advanced and recovery rates are low. Disturbance of underground aquifers and contamination of ground water are still major problems of the in situ process.

In 1972 the National Petroleum Council attempted to estimate costs.[29] Assuming a 15-percent rate of return on capital, the possible price of oil shale would be from $5.58 to $5.79 per barrel at a proportion of recovery available in extensive reserves of oil shale. For coal liquefaction, at a 15-percent rate of return on capital, per barrel costs are $7.75 to $8.25. These estimates indicate that oil shale would have a 38-to-42-percent cost advantage over coal liquefaction in the West and would probably be the source of first attempts to obtain oil from oil shale in the West. Again, the issue of whether the Western public would allow any large development of the oil-shale industry is important.

NOTES

1. Thadis Box et al., *Rehabilitation Potential of Western Coal Lands*, National Academy of Sciences (Cambridge: Ballinger Publishing Company, 1974).

2. As reported by Thadis Box, "Land Rehabilitation," *Coal Age,* May 1974, p. 109.

3. Ibid., pp. 109-10.

4. For a generally critical tone, see the voluminous United States Congress, Senate, Committee on Interior and Insular Affairs, Hearings, *Problems of Electrical Power Production in the Southwest,* Parts 1-5 (Washington, D.C.: Government Printing Office, 1971).

5. Indeed, this water availability resulted in the development of the major proportion of Arizona's coal production since a close market had now occurred. See Box, op. cit., p. 76.

6. Richard L. Gordon, *U.S. Coal and the Electric Power Industry* (Baltimore: Johns Hopkins University Press, 1975), Chapter 4, p. 71.

7. Ibid., p. 4.

8. See United States Bureau of Mines, *Strippable Reserves of Bituminous Coal and Lignite in the United States,* IC8531 (Washington, D.C.: Government Printing Office, 1971).

9. Ibid., p. 16.

10. Gordon, op. cit., Chapter 5, p. 26.

11. Massachusetts Institute of Technology, Energy Policy Study Group, "Energy Self-Sufficiency: An Economic Evaluation," *Technology Review*, May 1974: 39.

12. Federal Power Commission, Form 423 data, calendar year 1973 (Washington, D.C.: Government Printing Office, 1973).

13. The following discussion is based on Congressional Research Service, *Factors Affecting the Use of Coal in Present and Future Energy Markets* (Washington, D.C.: Government Printing Office, 1973), pp. 20-22.

14. "Environment," *Business Week*, August 3, 1974, p. 46.

15. Congressional Research Service, op. cit., p. 21.

16. United States Congress, Senate, Resolution 377, *Congressional Record*, October 12, 1972, pp. S17605-17607.

17. This is seen, for example, in Gordon, op. cit.

18. National Coal Association, *Bituminous Coal Data, 1973* (Washington, D.C.: National Coal Association, 1974), p. 82.

19. National Coal Association, *Bituminous Coal Facts, 1970* (Washington, D.C.: National Coal Association, 1971), p. 89.

20. Reported in Phillip Giffin, *Industrial Concentration and Firm Diversification in Bituminous Coal*, Appalachian Resources Project (Knoxville, Tennessee: The University of Tennessee, 1972), p. 45.

21. Ibid. We note that the single slurry-pipeline in Arizona is a result of coal extraction in an area where no railroad facilities exist, see *Coal Age*, May 1974, p. 77.

22. See the discussion on coal transport in Giffin, op. cit.

23. See United States Congress, Senate, Committee on Commerce, Hearings, *Freight Car Shortages*, 91st Congress, 2nd Session (Washington, D.C.: Government Printing Office, 1970).

24. Executive Office of the President, Office of Emergency Preparedness, *The Potential for Energy Conservation: Substitution for Scarce Fuels* (Washington, D.C.: Government Printing Office, 1973), pp. 48-51.

25. Conversation with Leonard Westerstrom, Division of Fossil Fuels, United States Bureau of Mines, Washington, D.C., 1975.

26. Energy Policy Project of the Ford Foundation, *Exploring Energy Choices* (Washington, D.C.: Ford Foundation, 1974), p. 36.

27. United States Congress, Senate, Committee on Commerce, op. cit.

28. Energy Policy Project of the Ford Foundation, op. cit., p. 27.

29. National Petroleum Council, Committee on U.S. Energy, *U.S. Energy Outlook*, December 1972, Chapter 5.

CHAPTER 9

Coal use in conventional steam electric generation in the Eastern and Midwestern markets, though subject to environmental pressures to reduce sulfur emissions does not show any prospect of decreasing in importance in the near future. Table 9.1 shows the planned capacity of new generating units by type. Over 60 percent of the additions to conventional steam electric generation are coal-fired units. We have seen that consumption of electricity has increased faster than overall energy demand in the past, and coal will continue to be the main generating source for electricity.

Table 9.2 shows the average prices paid by utilities for the three fossil fuels in the Eastern states. The competitive position of coal at present is fairly strong in most states. Use of natural gas as a utility fuel has been constrained by regulations. The four states of Texas, Louisiana, Oklahoma, and New Mexico produce almost 90 percent of U.S. natural gas and more than 95 percent of the net gas delivered in interstate commerce.[1] Interstate sales are those in which the Federal Power Commission regulates the wellhead prices. The prospects for complete deregulation of gas may well be politically nil. It has been suggested that even if new supplies were priced at 80 cents or more per million cubic feet (MCF), a gap between available supply and prospective demand would exist.[2] This rate would be more than three times the present average for interstate sales. Stauffer and Jensen indicate that a wellhead parity price of $1.20/MCF, almost six times the present interstate average, is possible under complete deregulation.[3]

We have seen that natural gas is used in electric generation in the Southwest Central region since intrastate sales are not regulated and higher prices can be bid for local natural-gas production. Before natural gas can be deregulated, its possible resale must be considered, since utilities in this area have large future supplies tied up in long-term contracts.[4] Prospective supplies of natural gas and

TABLE 9.1

Type of Generating Units as a Percent of Total Annual New Capacity in the United States, Excluding Southwest Central and Pacific Regions

Type of Unit	1972	1973	1974	1975	1976
All conventional steam electric units (coal, oil, and gas)	52.3	44.7	49.2	68.2	72.4
Coal-fired	40.3	36.1	31.0	43.5	42.5
Nuclear	29.5	37.4	41.3	23.7	18.7
Other	18.2	17.9	9.5	8.1	8.9
Total	100.0	100.0	100.0	100.0	100.0

Note: Conventional plants can be "on line" some four years after the order is placed. Atomic power plants can require as much as eight years. Therefore, any comparison of conventional versus nuclear capacity for 1976 and onward would be distorted. The figure for conventional plants can increase because, at the time these data were collected, they could still be ordered for service in 1976 and thereafter; atomic plants cannot increase until 1978 and thereafter.

Source: National Coal Association, *Steam Electric Plant Factors, 1972* (Washington, D.C.: National Coal Association, 1972), p. 54.

their limitations, even at the higher prices, indicate that natural gas under deregulation would not be a competitive utility fuel.[5]

Oil prices are subject to the availability and price of imported supplies, particularly from the Middle East. More recently, we have seen that traditionally friendly countries, such as Canada, worried about their own future energy supplies, are considering phasing out all of their oil exports to the United States. In the Midwest and Eastern states coal costs increased most because of the ban on steep-slope surface mining and the high tax rate on sulfur emissions. However, the average increase of 65 cents per million Btu would still leave coal priced favorably to gas in the future. Whether the use of oil would increase significantly would depend on its availability as well as its cost. Perhaps more important, even if reclamation policies increase production costs as much as $2.00 per ton, it would only mean an increase of 8-9 cents per million Btu. This additional cost, given the figures in Table 9.2, leads to a tentative conclusion that the effects of reclamation requirements on the position of coal as a boiler fuel would be minimal, particularly in existing plants where the cost of converting boilers from coal to oil must be considered.

TABLE 9.2

Unit Costs of Fossil Fuels Used in Electric Generation in the East and Midwest

Type of Fuel	Year	New England	Middle Atlantic	Northeast Central	Northwest Central	South Atlantic	Southeast Central	Southwest Central
Coal	1969	36.9	30.0	26.4	26.2	28.4	21.1	*
	1970	41.9	36.1	30.4	28.2	36.1	23.6	*
	1971	48.8	40.9	35.5	31.6	41.8	29.2	17.8
	1972	49.7	42.1	38.9	34.0	42.6	32.5	21.0
	1973	52.8	46.3	41.8	35.5	44.9	37.7	12.8
Percent increase	1972/71	+1.8	+2.9	+9.6	+7.6	+1.9	+11.3	+18.0
	1973/72	+6.2	+10.0	+7.4	+4.4	+5.4	+16.0	-60.0
Oil	1969	28.3	33.6	62.0	51.8	30.4	51.1	36.9
	1970	32.8	40.2	56.7	59.0	31.9	54.1	44.6
	1971	47.6	57.1	63.2	70.3	43.3	49.6	59.8
	1972	55.5	62.3	68.0	69.9	49.6	72.4	67.2
	1973	74.9	82.4	78.9	90.4	67.6	96.8	97.5
Percent increase	1972/71	+16.6	+9.1	+7.6	-0.6	+14.5	+46.0	+12.4
	1973/72	+34.9	+32.3	+16.0	+29.3	+36.3	+33.7	+45.1
Gas	1969	33.7	35.6	31.6	24.9	31.6	24.3	20.5
	1970	35.3	38.3	37.1	25.6	34.7	25.3	21.1
	1971	45.5	44.9	42.9	28.3	39.7	27.9	22.2
	1972	46.1	53.1	51.6	29.9	39.9	29.9	24.2
	1973	52.5	50.8	55.2	35.1	45.2	37.7	28.1
Percent increase	1972/71	+1.3	+18.3	+20.3	+5.7	+0.5	+7.2	+9.0
	1973/72	+13.9	-3.0	+7.0	+19.4	+13.3	+26.1	-16.1

*Not shown since quantities consumed were negligible in data sources.

Note: Figures are "as Burned Cost," per million Btu, cents.

Source: Data until 1972 from National Coal Association, *Steam Electric Plant Factors, 1972* (Washington, D.C.: National Coal Association, 1973), p. 54; 1973 data from Federal Power Commission, *Monthly Report of Cost and Quality of Fuels for Steam Electric Plants,* December 1973 (Washington, D.C.: Government Printing Office, 1974), p. 22.

TABLE 9.3

Known and Recoverable Coal Reserves in the Eastern Fields
(million short tons)

| | | | Regional* Reserves | | | | | |
| Sulfur content (by weight, dry basis) | Northern Appalachian | | Southern Appalachian | | Eastern Interior | | Western Interior | |
	Known	Rec.	Known	Rec.	Known	Rec.	Known	Rec.
≤ 0.7	45	5	37,275	4,100	195	50	250	20
0.8-1.0	2,755	360	41,025	4,510	1,355	340	770	60
1.1-1.5	21,370	2,780	18,135	1,995	10,075	2,520	2,475	200
1.6-2.0	23,050	2,995	9,890	1,090	9,135	2,285	1,180	95
2.1-2.5	27,525	3,580	2,770	305	7,440	1,860	9,170	735
2.6-3.0	11,950	1,550	3,510	385	32,300	8,075	2,070	165
3.1-3.5	8,780	1,140	275	30	67,065	16,765	11,340	905
3.6-4.0	7,155	930	45	5	86,755	21,690	28,975	2,320
> 4.0	800	105	85	10	35,525	8,890	62,685	5,015
Total	103,340	13,445	113,010	12,430	249,845	62,465	118,915	9,515
Percent of total reserves recoverable	13		11		25		8	

Note: Total Reserves 585,199
Total Recoverable Reserves 97,855
Total Strippable Reserves 70,621

*Northern Applachian is Pennsylvania, northern West Virginia, and Maryland; Southern Appalachian is eastern Kentucky, southern West Virginia, Tennessee, Virginia, and Alabama; Eastern Interior is Illinois, Indiana, western Kentucky and Ohio; Western Interior is Iowa, Kansas, Missouri, Oklahoma, Arkansas, and Texas.

Source: Mitre Corporation, *Survey of Coal Availability by Sulfur Content*, Environmental Protection Agency (Washington, D.C.: Government Printing Office, 1972); for total strippable reserves: United States Bureau of Mines, *Strippable Reserves of Bituminous and Lignite in the United States*, IC8531 (Washington, D.C.: Government Printing Office, 1971).

The availability of Eastern coal reserves is seen in Table 9.3. The Bureau of Mines study used in considering Western coal reserves indicates that approximately 70,621 million tons of known Eastern reserves are strippable, but only 18,372 million tons can be recovered under present mining conditions and costs.[6] This recovery percentage of 26 percent is less than half the Western figure of 56 percent. At an annual production level of 800 million tons, for example, these strippable reserves would last 23 years. The remaining reserves, 79,483 million tons, are recoverable only by underground mining techniques. In contrast to the strippable reserves, at a production level of 800 million tons, Eastern underground reserves would last almost 100 years.

The majority of the lower-sulfur reserves in Appalachia can be obtained by underground mining. Even if we were to assume that all recoverable and strippable reserves in Appalachia contained less than 1.5 percent sulfur, they would comprise only 47 percent of the total of almost 14 billion tons at low-sulfur levels. This represents a significant level of low-sulfur reserves, considering that the figures are based on deposits mineable under existing conditions. Even at this level, however, low-sulfur deposits in the East could provide an average of 350 million tons for 40 years. In the Midwest, the recoverable resources of coal with less than 1.5 percent sulfur content are comparatively low, that is, 3.19 billion tons. Most of the Midwestern reserves, however, may be recovered by surface mining. Again, these figures do not include additional reserves that might become recoverable as economic incentives to utilize them develop.

How sulfur emissions standards would affect the use of Western coal in Midwestern utility markets was clearly seen earlier in this study. Table 9.4 shows the effects of alternative levels of sulfur emissions standards for electric utility power plants. The limited availability of lower-sulfur coal from non-Western producers during a single production year is the "bottleneck" which causes these results. The increase in Western production as the allowable level of sulfur emissions is decreased compensates for Midwestern coal which can no longer be utilized. The potential for Midwestern coal deposits in the future is affected by the degree to which sulfur emissions are limited by public policy. We have seen that the Midwestern coal fields are a lowest cost source of supply for utilities in states in that area. Our results indicate that an increased reliance on coal resulting from a policy of domestic self-sufficiency in energy for the near future would require a decision between controlling sulfur emissions and establishing a large Western coal industry. Increases in Western surface mines could well take less than half the lead time of that needed for the development of low-sulfur underground-mine expansion in Appalachia.

The prospects for any large dependence on Eastern underground mining are not clear. Traditionally underground mining has been associated with the economic development of many rural Appalachian counties. Studies have shown that increase in population, regional income, and the level of trade are linked to

TABLE 9.4

Regional Shipments with Alternative Sulfur Emissions Standards (millions of tons)

Region	Sulfur Emissions Standard, pounds per million Btu			
	None	2.00	1.75	1.65
Appalachia	307.9	325.6	310.2	302.2
Midwest	154.0	149.2	117.8	106.5
West	41.9	49.0	81.5	105.5
United States	503.8	523.8	509.5	501.6
Sulfur Emissions, millions of tons	11.41	10.64	9.80	9.30

Source: Compiled by author.

deep-mining development.[7] Changes in coal demand can produce "boom-or-bust" tendencies in such areas but do not cause the excessive out-migration and population decline seen in strip-mining areas. We have seen that the adequacy of labor supply for underground expansion, from mining engineers to miners themselves, is a cause for concern. Since underground production requires twice the labor input of surface mining (and more highly skilled input at that) any large-scale development of underground mining would be delayed by the labor constraint alone. Possible control by a single union, the United Mine Workers, of an underground labor force twice its present size could result in large cost increases and a strike could interrupt the nation's energy supply far more than any Arab oil embargo.

Future technology could have two important effects on Eastern and Midwestern high-sulfur coal deposits. Coal gasification could make these higher-sulfur coals almost sulfur-free. Unfortunately, it is difficult to use the high-coking coals of the Eastern fields in the most advanced technique at present available for gasification, the Lurgi process.[8] Assuming a 15-percent cost of capital and coal at 40-50 cents per million Btu, the cost estimates for gasification would range from $1.63 to $1.70 per million Btu.[9] Changes in the price of coal utilized for gasification are significant because the efficiency of conversion is only 67 percent. Thus a rise in coal price of 8 cents per million Btu would raise coal gasification costs by 12 cents per million Btu. Because of this conversion factor, any long-term reliance on gasification would deplete coal reserves at a faster rate than if they were mined and used directly.

The second method which could alter the prospects of higher-sulfur coals is stack-gas scrubbing. Having reviewed the current technology, lead times, and problems, let us concentrate on the cost implications. It is of interest to determine the cost levels at which the use of local high-sulfur coal would be cheaper than that of Western coal, particularly in the Midwestern markets. In our study, the delivered cost of Western coal in Chicago is 62 cents per million Btu.[10] At this delivered cost level, in order for local Illinois production to be low-sulfur and competitive with Western coal, the allowable cost of scrubbing is around 22-24 cents per million Btu. Cost estimates for scrubbing systems, when they become available, are much higher. The midpoint of the cost range is 58 cents per million Btu.[11] Research by Commonwealth Edison indicates that plant-utilization rates fall in later years of operation. The cost range of 50 to 63 cents per million Btu will rise to 75 to 85 cents in the later years of a plant's life.[12] Existing plants attempting to retrofit scrubbers will probably face these higher figures also.

At these cost levels, Western coal is cheaper, even with reclamation requirements. Even at the lowest estimate of 50 cents, it would be cheaper for Michigan utilities to use Western coal than to use Ohio production and employ scrubbers. The reserves of low-cost strippable Western coal are sufficient to supply much of the Midwestern markets if extensive development is encouraged. Furthermore, Western producers could be quite competitive with coal gasification in the Midwest. In the markets traditionally served by Appalachian producers, however, such as Pennsylvania, New York, etc., utilities would find it as much as $7 per ton cheaper to turn to underground low-sulfur production rather than use Western coal.

NOTES

1. James Hensen and Thomas Stauffer, "An Economic Rationale for Rationing Gas Supplies in the U.S." (Cambridge: Arthur D. Little, Inc., 1974), p. 3.

2. James Hensen and Thomas Stauffer, "The Rational Allocation of Natural Gas Under Chronic Supply Constraints," in *Energy: Demand, Conservation and Institutional Problems* (Cambridge: Massachusetts Institute of Technology Press, 1974), p. 289.

3. Ibid.

4. Richard L. Gordon, *U.S. Coal and the Electric Power Industry* (Baltimore: Johns Hopkins University Press, 1975), Chapter 5, p. 93.

5. See for example, Gordon, op. cit., Chapter 5, pp. 10-12.

6. See United States Bureau of Mines, *Strippable Reserves of Bituminous Coal and Lignite in the United States*, IC8531 (Washington, D.C.: Government Printing Office, 1971).

7. United States Congress, House, Committee on Interior and Insular Affairs, Hearings, *Regulation of Surface Mining,* Part I (Washington, D.C.: Government Printing Office, 1973), p. 780.

8. K. Hottel and J. Howard, *New Energy Technology* (Cambridge: Massachusetts Institute of Technology Press, 1971), p. 146.

9. These represent updated estimates with higher coal prices of National Petroleum Council figures as reported in Gordon, op. cit., Chapter 5, p. 17.

10. Our figures are in agreement with those, for example, in Massachusetts Institute of Technology, Energy Policy Study Group, "Energy Self-Sufficiency: An Economic Evaluation," *Technology Review,* May 1974, p. 40.

11. Ibid., p. 41.

12. Reported in Richard L. Gordon, "Environmental Impacts in Energy Production and Use," "Energy Supply Project" (unpublished manuscript on file, Resources for the Future, Washington, D.C., 1973), p. 123. This section summarizes the literature on cost estimates.

BIBLIOGRAPHY

Abrams, Lawrence and Barr, James. "Corrective Taxes for Pollution Control." *Journal of Environmental Economics and Management* 1 (1974): 296-318.

Arrow, Kenneth, *Essays in the Theory of Risk Bearing*. Chicago: Markham Publishing Company, 1971.

Appalachian Regional Commission. *Acid Mine Drainage in Appalachia*. H. R. Doc. No. 90-180, 91st Congress, 1st Session, XXIV. Washington, D.C.: Government Printing Office, 1969.

Averitt, Paul. *Stripping Coal Resources of the United States—January 1, 1970*. U.S. Geological Survey, Bulletin 1322, Washington, D.C: Government Printing Office, 1970.

Barnett, Harold. *Energy, Resources, and Growth*. Department of Economics, Washington University, St. Louis, Missouri, 1973.

Barrett, L., and Waddell, T. *Costs of Air Pollution Damages: A Status Report*. Research Triangle Park, North Carolina: Environmental Protection Agency, 1973.

Baumol, W., and Oates, W. "The Use of Standards and Pricing for the Protection of the Environment." *Swedish Journal of Economics* 73 (1971): 42-54.

Bish, Robert. *The Public Economy of Metropolitan Areas*. Markham Public Finance Series. Chicago: Markham Publishing Company, 1971.

Bohm, Robert, et al. *Benefits and Costs of Surface Coal Mining in Appalachia*. Appalachian Resources Project. Knoxville, Tennessee: The University of Tennessee, 1971.

Box, Thadis. "Land Rehabilitation." *Coal Age*. May 1974, pp. 109-110.

Box, Thadis, et al. *Rehabilitation Potential of Western Coal Lands*. National Academy of Sciences. Cambridge, Massachusetts: Ballinger Publishing Company, 1974.

Business Week. "Environment." August 3, 1974, p. 46.

Business World. August 3, 1974.

Chapman, Duane. "Internalizing an Externality: A Sulfur Emission Tax and the Electric Utility Industry." *Energy: Demand, Conservation, and Institutional Problems*. Cambridge: Massachusetts Institute of Technology Press, 1974, pp. 190-208.

Chapman, Duane, et al. "Electricity and the Environment." Paper presented at AAAS meeting, Philadelphia, December 26, 1971.

————. "Electricity Demand Growth and the Energy Crisis." *Science*, 178, no. 4062 (November 1972): 703-8.

Charles River Associates. *The Economic Impact of Public Policy on the Appalachian Coal Industry and the Regional Economy.* Vols. 1 and 2. Cambridge, Massachusetts: Charles River Associates, 1973.

Coal Age. "New Look at Western Coal." May 1974, pp. 75-130, also see pp. 9, 27, and 91.

Coal Age. "Western Coal Development." May 1975, p. 90.

Collier, C. R., et al. *Influences of Strip Mining on the Hydrologic Environment of Parts of Beaver Creek Basin, Kentucky, 1955-66.* United States Department of Interior. Washington, D.C.: Government Printing Office, 1970.

Committee on U.S. Energy. *U.S. Energy Outlook.* National Petroleum Council, Washington, D.C., 1972.

Congressional Research Service. *Factors Affecting the Use of Coal in Present and Future Energy Markets.* Washington, D.C.: Government Printing Office, 1973.

Council on Environmental Quality. *Coal Surface Mining and Reclamation.* Washington, D.C.: Government Printing Office, 1973.

————. *Quantitative Energy Studies and Models.* Washington, D.C.: Government Printing Office, 1973.

————. *The President's 1972 Environmental Program.* Washington, D.C.: Government Printing Office, 1972.

Culbertson, Oran L. *The Consumption of Electricity in the United States.* Oak Ridge, Tennessee: Oak Ridge National Laboratories, June 1971.

Curry, Robert. "Reclamation Considerations for the Arid Lands of the Western United States." Statement prepared for United States Congress, House, Committee on Interior and Insular Affairs. Published in *Regulation of Surface Mining Operations,* Part 2. Washington, D.C.: Government Printing Office, 1973, pp. 1006-15.

Denis, Sylvian. *Some Aspects of the Environment and Electric Power Generation.* Santa Monica, California: Rand Corporation, 1972.

Dorfman, Robert, et al. *Linear Programming and Economic Analysis.* New York: McGraw-Hill Book Company, Inc., 1958.

Edison Electric Institute. *Statistical Yearbook.* New York: Edison Electric Institute, 1973.

Electrical World. "Twenty-Third Annual Electrical Industry Forecast." New York, 1972.

Energy Policy Project of the Ford Foundation. *Exploring Energy Choices.* Washington, D.C.: Ford Foundation, 1974.

Environmental Policy Center. *Environment.* Washington, D.C.: Environmental Policy Center, August 1974.

Executive Office of the President, Office of Science and Technology, Energy Policy Staff. *Considerations Affecting Steam Power Plant Site Selection.* Washington, D.C.: Government Printing Office, 1970.

————. Office of Emergency Preparedness. *The Potential for Energy Conservation; Substitution for Scarce Fuels.* Washington, D.C.: Government Printing Office, 1973.

Federal Power Commission. *Monthly Report of Cost and Quality of Fuels for Steam Electric Plants.* (All issues.) Washington, D.C.: Government Printing Office, 1972-74.

————. *Problems in Disposal of Waste Heat from Steam Electric Plants.* Washington, D.C.: Government Printing Office, 1969.

Giffin, Phillip. *Industrial Concentration and Firm Diversification in Bituminous Coal.* Appalachian Resources Project. Knoxville, Tennessee: The University of Tennessee, 1972.

Goldberg, Everett, et al. *Legal Problems of Coal Mine Reclamation.* Environmental Protection Agency. Washington, D.C.: Government Printing Office, 1972.

Gordon, Richard L. *U.S. Coal and the Electric Power Industry.* Baltimore: Johns Hopkins University Press, 1975.

————. "Coal—Our Limited Vast Fuel Resource," in E. Erikson, et. al., *The Energy Question,* Vol. 2. Toronto: University of Toronto Press, 1974, pp. 49-75.

————. "Environmental Impacts of Energy Production and Use." "Energy Supply Project." Unpublished manuscript on file, Resources for the Future, Washington, D.C., 1973.

Griffin, James. "Recent Sulfur Tax Proposals: An Econometric Evaluation of Welfare Gains." In *Energy: Demand, Conservation, & Institutional Problems,* edited by Michael S. Macrakis. Cambridge: Massachusetts Institute of Technology Press, 1974, pp. 236-48.

Hagen, Hubert. "Regarding Proposed Surface Mine Legislation." Statement prepared for United States Congress, Senate, Committee on Interior and Insular Affairs, Hearings, 93rd Congress, 1st Session. Published in *Regulation of Surface Mining Operations,* Parts 1 and 2. Washington, D.C.: Government Printing Office, 1973.

Hagevik, George. *Decision Making in Air Pollution Control.* New York: Praeger Publishers, 1970.

Hausgaard, Q. "Proposed Tax on Sulfur Content of Fossil Fuels." *Public Utilities Fortnightly* 88 (6): 27-33.

Henderson, James. *The Efficiency of the Coal Industry.* Cambridge, Massachusetts: Harvard University Press, 1958.

————. "A Short Run Model of the Coal Industry." *Review of Economics and Statistics* 37 (1955): 334-47.

Hensen, James, and Stauffer, Thomas. "An Economic Rationale for Rationing Gas Supplies in the U.S." Cambridge, Massachusetts: Arthur D. Little, Inc., 1974.

————. The Rational Allocation of Natural Gas Under Chronic Supply Constraints." In *Energy: Demand, Conservation, and Institutional Problems*, edited by Michael S. Macrakis. Cambridge: Massachusetts Institute of Technology Press, 1974.

Hittman Associates, Inc. *Assessment of 502 Control Alternatives and Implementation Patterns for the Electric Utility Industry*. Environmental Protection Agency. Washington, D.C.: Government Printing Office, 1973.

Hottel, K., and Howard, J. *New Energy Technology*. Cambridge, Massachusetts Institute of Technology Press, 1971.

Howard, Herbert A. "External Diseconomies of Bituminous Coal Surface Minging—A Case Study of Eastern Kentucky." Unpublished doctoral dissertation, Indiana University, Bloomington, Indiana, 1968.

Jimeson, Robert M., and Chilton, Cecil. "A Model for Determining the Minimum Cost Allocation of Fossil Fuels." Paper presented to the Sociedad Mexicana de Ingenieria de Costos, Second National Cost Engineering Congress, Mexico City, October 29-November 1, 1972.

Joint Committee on Atomic Energy. *Nuclear Power and Related Energy Problems: 1968 Through 1970*. Washington, D.C.: Government Printing Office, 1971.

Kneese, A., and Bower, B. *Managing Water Quality: Economics, Technology, and Institutions*. Baltimore: Johns Hopkins Press, 1968.

Lave, Lester. "Air Pollution Damages: Some Difficulties in Estimating the Value of Abatement," in *Environmental Quality Analysis*, A. Kneese et al., eds. Baltimore: Johns Hopkins Press, 1972.

Lefeber, Louis. *Allocation in Space: Production, Transport, and Industrial Location*. Amsterdam: North-Holland, 1958.

Massachusetts Institute of Technology, Energy Policy Group. "Energy Self-Sufficiency: An Economic Evaluation." *Technology Review*, May 1974, pp. 38, 39, 42-43.

Miernyk, William. "Environmental Management and Regional Economic Development." Paper presented at the Southern Economic Association meeting, Miami Beach, Florida, November 6, 1971.

Mined Land Conservation Conference. *What About Strip Mining*. Washington, D.C.: Mined Land Conservation Committee, 1964.

Mitre Corporation. *Survey of Coal Availabilities by Sulfur Content*. Environmental Protection Agency. Washington, D.C.: Government Printing Office, 1972.

Moses, Leon. "The General Equilibrium Approach," in Robert Dean et al., eds., *Spatial Economic Theory*. New York: Free Press, 1970.

Mount, T., et al., *Electricity Demand in the United States: An Econometric Analysis*. Oak Ridge, Tennessee: Oak Ridge National Laboratories.

Moyer, Reed. *Competition in Midwestern Coal Industry*. Cambridge, Massachusetts: Harvard University Press, 1964.

National Air Pollution Control Administration. *Control Technique for Sulfur Oxide Pollutants*. Washington, D.C.: Government Printing Office, 1969.

National Coal Association. *Bituminous Coal Data, 1973*. Washington, D.C.: National Coal Association, 1974.

———. *Bituminous Coal Facts, 1970*. Washington, D.C.: National Coal Association, 1971.

———. *Steam Electric Plant Factors, 1972*. Washington, D.C.: National Coal Association, 1973.

National Petroleum Council, Committee on U.S. Energy. *U.S. Energy Outlook*. Washington, D.C.: December 1972.

National Safety Council. *Accident Facts 1971*. Washington, D.C.: National Safety Council.

Newsweek. August 5, 1974, cited in *Environment*. Washington, D.C.: Environmental Policy Center, August, 1974.

Nordhaus, William. "The Allocation of Energy Reserves." Paper presented at Brookings Panel on Economic Activity, Washington, D.C., November 15, 1973.

Resources for the Future. "Energy Supply Project." Unpublished manuscript on file, Washington, D.C., 1973.

Ridker, R. *The Economic Costs of Air Pollution*. New York: Praeger Publishers, 1967.

Ridker, R., and Henning, J. "The Determinants of Residential Property Values with Specific Reference to Air Pollution." *Review of Economics and Statistics* 49 (1967): 246-57.

St. Louis *Post Dispatch*. September 25, 1974.

Samuelson, Paul. *The Collected Scientific Papers of Paul Samuelson*. Vol. 2. Cambridge: Massachusetts Institute of Technology Press, 1966.

Schmidt-Bleek, F. K., et al. *Benefit-Cost Evaluation of Strip Mining in Appalachia*. Appalachian Resources Project. Knoxville, Tennessee: The University of Tennessee, 1973.

Spore, Robert. "The Economic Problems of Coal Surface Mining." *Environmental Affairs* 2, no. 4 (June, 1973): 585-593.

Spore, Robert, et al. "Opportunity Costs of Land Use: The Case of Coal Surface Mining." *Energy-Demand, Conservation, and Institutional Problems*. Cambridge: Massachusetts Institute of Technology Press, 1973.

Sporn, Phillip, ed. *The Tall Stack for Air Pollution Control on Large Fossil Fuel Power Plants*. Washington, D.C.: National Coal Policy Conference, Inc. 1967.

Sulfur Oxide Control Technology Assessment Panel. *Final Report on Projected Utilization of Stack Gas Cleaning Systems by Steam Electric Plants*. Washington, D.C.: Government Printing Office, 1973.

Tabb, William. *A Recursive Programming Model of Resource Allocation*. Unpublished doctoral dissertation, University of Wisconsin, Madison, Wisconsin, 1968.

Tennessee Valley Authority. *A Quality Environment in the Tennessee Valley*. Knoxville, Tennessee: Tennessee Valley Authority, 1974.

———. *TVA Today 1974*. Knoxville, Tennessee: Tennessee Valley Authority, 1974.

Train, Russell. "Coal Burning Power and the E.P.A." Washington *Post*, September 5, 1974.

United States Bureau of Mines. *Basic Estimated Capital Investment and Operating Costs for Underground Bituminous Coal Mines.* IC 8632, Washington, D.C.: Government Printing Office, 1974.

―――. *Bitumous Coal and Lignite Distribution 1973.* Washington, D.C. Government Printing Office, 1973.

―――. *Bituminous Coal and Lignite Mine Openings and Closings in the Continental United States, 1970, 1971, 1972.* Washington, D.C.: Government Printing Office, 1973.

―――. *Coal–Bituminous and Lignite in 1972.* Washington, D.C.: Government Printing Office, 1973.

―――. *Coal Production from the Unita Region, Colorado and Utah–Cost Analysis for Proposed Underground Mining Operations.* IC 8497. Washington, D.C.: Government Printing Office, 1970.

―――. *Cost Analyses of Model Mines for Strip Mining of Coal in the United States.* IC 8535. Washington, D.C.: Government Printing Office, 1972.

―――. *Land Utilization and Reclamation in the Mining Industry, 1930-1971.* IC 8642. Washington, D.C. Government Printing Office, 1974.

―――. *Mineral Yearbook, 1973.* Washington, D.C.: Government Printing Office, 1974.

―――. *Strippable Reserves of Bitminous Coal and Lignite in the United States.* IC 8531. Washington, D.C.: Government Printing Office, 1971.

United States Congress. House, Committee on Interior and Insular Affairs. Hearings. Published in *Regulation of Surface Mining.* Part I. Washington, D.C.: Government Printing Office, 1973.

―――. Senate, Resolution 377. *Congressional Record*, October 12, 1972, S17605-17607. Washington, D.C.: Government Printing Office, 1972.

―――. Senate. Statement by Senator William Proxmire, 92nd Congress, 2nd Session. *Congressional Record*, 118(5), pp. 276-79. Washington, D.C.: Government Printing Office.

―――. Senate, Committee on Commerce, Hearings. *Freight Car Shortages.* 91st Congress, 2nd Session, Washington, D.C.: Government Printing Office, 1970.

―――. Senate, Committee on Interior and Insular Affairs, Hearings. *Problems of Electrical Power Production in the Southwest*, Parts 1-5, Washington, D.C.: Government Printing Office, 1971.

United States Department of Interior. *Surface Mining and Our Environment.* Washington, D.C..: Government Printing Office, 1967.

―――. *The Administration of the Federal Coal Mine Health and Safety Act.* Washington, D.C.: Government Printing Office, 1974.

United States Environmental Protection Agency. *Compilation of Air Pollutant Emission Factors*, Research Triangle Park, North Carolina: Environmental Protection Agency, 1973.

————. *Nationwide Air Pollutant Emission Trends 1940-1970.* Research Triangle Park, North Carolina: Environmental Protection Agency, 1973.

————. Office of Research and Development. *Environmental Considerations in Future Energy Growth.* Vol. I, *Fuel-Energy Systems.* Washington, D.C.: Government Printing Office, 1973.

ALAN M. SCHLOTTMANN is Assistant Professor of Economics at the University of Tennessee, Knoxville. He has worked extensively in the area of energy policy and the natural resource development of Appalachia.

Professor Schlottmann's research at the Appalachian Resources Project of the University of Tennessee has focused on alternative energy and environmental policies relating to coal use, thus laying the foundation for the present study. Dr. Schlottmann received his Ph.D. from Washington University, St. Louis, and has written articles for such professional journals as *Land Economics, Review of Economics and Statistics,* and *Journal of Regional Science.*

RELATED TITLES
Published by
Praeger Special Studies

ALTERNATIVE ENERGY STRATEGIES: Constraints
and Opportunities
by John Hagel III

ENVIRONMENTAL POLITICS
edited by Stuart S. Nagel

THE DYNAMICS OF ELECTRICAL ENERGY SUPPLY
AND DEMAND: An Economic Analysis
by R. K. Pachauri

ENVIRONMENTAL LEGISLATION: A Sourcebook
edited by Mary Robinson Sive